T4-ANG-193

When John Maynard Keynes likened Jan Tinbergen's early work in econometrics to black magic and alchemy, he was expressing a widely held view of a new discipline. However, even after half a century of practical work and theorizing by some of the most accomplished social scientists, Keynes' comments are still repeated today.

This book assesses the foundations and development of econometrics and sets out a basis for the reconstruction of the foundations of econometric inference by examining the various interpretations of probability theory which underlie econometrics. Keuzenkamp contends that the probabilistic foundations of econometrics are weak, and, although econometric inferences may yield interesting knowledge, claims to be able to falsify or verify economic theories are unwarranted. Methodological falsificationism in econometrics is an illusion. Instead, it is argued, econometrics should locate itself in the tradition of positivism.

HUGO KEUZENKAMP is professor in applied economics at the University of Amsterdam and director of the Foundation for Economic Research in Amsterdam. Until recently he was the editor-in-chief of *Economisch Statistische Berichten*, a Dutch weekly periodical on economic policy. He has published widely in journals, including *Economic Journal*, *Journal of Econometrics* and *Journal of Economic Surveys*.

Probability, Econometrics and Truth

The methodology of econometrics

Hugo A. Keuzenkamp

PUBLISHED BY THE PRESS SYNDICATE OF THE UNIVERSITY OF CAMBRIDGE
The Pitt Building, Trumpington Street, Cambridge, United Kingdom

CAMBRIDGE UNIVERSITY PRESS
The Edinburgh Building, Cambridge CB2 2RU, UK www.cup.cam.ac.uk
40 West 20th Street, New York, NY 10011-4211, USA www.cup.org
10 Stamford Road, Oakleigh, Melbourne 3166, Australia
Ruiz de Alarcón 13, 28014 Madrid, Spain

First published 2000

Printed in the United Kingdom at the University Press, Cambridge

Typeface 10/12 Times *System* 3B2 [KW]

A catalogue record for this book is available from the British Library

Library of Congress Cataloguing in Publication data

Keuzenkamp, Hugo.
Probability, econometrics and truth: the methodology of econometrics/Hugo Keuzenkamp.
p. cm.
Includes indexes.
ISBN 0 521 55559 8 (hardback)
1. Econometrics. 2. Probabilities. I. Title

HB139.K48 2000
330′.01′5195–dc21 99-058732

ISBN 0 521 553 59 8 hardback

Contents

Introduction

Probability begins and ends with probability.

Keynes ([1921] CW VIII, p. 356)

When John Maynard Keynes accused Jan Tinbergen of practising black magic and alchemy, econometrics was still in its infancy. A critical attitude to econometrics was legitimate, as it would have been for any novel enterprise. Stubborn perseverance on behalf of the pioneers of econometrics is natural as well. However, after more than half a century of development by some of the most brilliant social scientists, and much practical experience, Keynes' comments are repeated today by respected authorities. Has it all been in vain?

Not quite. It is true that the aspirations (or pretences) of econometrics and the accomplishments still tend to be divided by a gap, which, in turn, tends to damage the credibility of the whole discipline. Many of econometrics' results remain controversial. Some critics claim that even the most basic aim, the measurement and quantitative description of economic phenomena, has not been accomplished. Econometric evidence has been compared with the evidence of miracles in Lourdes. Some deplore the waste of electricity used for econometric computer calculations. But a fair appraisal of contemporary econometrics cannot deny that a number of interesting empirical lessons have been learnt. The verdict that the econometric exploration was all in vain can only result from a wrong interpretation of econometric aims.

This book is a methodological investigation of this exploration. It confronts the aims with the methods and with the philosophical as well as the probabilistic foundations of those methods. It concludes that the achievements of econometrics can be found where its aspirations are put in the positivist tradition. Positivism is a philosophy which has been declared dead by many. It should be resurrected.

Positivism has an ancestor in David Hume, one of the founders of British empiricism (the forerunner of positivism). Hume ([1748] 1977, p. 114) encouraged his readers to ask about any book in their libraries,

Does it contain any abstract reasoning concerning quantity or number? No. *Does it contain any experimental reasoning concerning matter of fact and existence?* No. Commit it then to the flames: For it can contain nothing but sophistry and illusion.

'Science is Measurement', the original motto of the Cowles Commission for Research in Economics (which had a leading role in shaping formal econometrics), put econometrics clearly in the positivist tradition. Twenty years later, in 1952, this motto was changed to 'Theory and Measurement', reflecting the ambitions of a younger generation of researchers (headed by Trygve Haavelmo and Tjalling Koopmans) to integrate econometrics with (neoclassical) economic theory and formal probability theory. The new tradition diverted econometrics to Neyman–Pearson testing procedures, away from the positivist tradition of Karl Pearson (father of Neyman's companion Egon), Fisher and Jeffreys. Simultaneously, in the philosophy of science positivism came under attack and was replaced by methodological falsificationism. Chapter 1 discusses this philosophy, chapters 2 to 4 deal with different approaches in probability theory. I claim that the Cowles programme in econometrics, with its Neyman–Pearson foundation and a philosophical sauce of methodological falsificationism, has done the reputation of econometrics much harm. This claim is elaborated in the chapters which follow the discussion of the various probability theories. The transition from probability theory to econometrics is shaky, as chapters 6 and 7 demonstrate. Chapter 8, which presents a case study in one of the best episodes of applied econometric inference, shows that the sampling and testing metaphors which dominated econometrics can lead to serious self-deceit. Chapters 9 and 10 bring various arguments together and recommend the positivist tradition, in which econometrics was born and to which it should be brought back again.

Finally, what about truth? Does econometrics, based on the right kind of probability, yield 'true knowledge'? Not so. The quest for truth, which dominates much of contemporary econometrics, should be abandoned. If econometricians are able to deliver useful approximations to empirical data, they achieve a major accomplishment. What defines 'useful' is an intricate matter, which can only be clarified on a case-by-case basis. A model which is good for one purpose, may be inappropriate for another.

I hope that the reader will allow the author a few disclaimers. First, the focus on econometrics does not mean that there are no other ways of doing empirical economics, neither is it intended to suggest that purely theoretical work is not interesting. This book intends to provide a complement to books on economic methodology which tend to ignore the strengths but also weaknesses of econometric inference.

Secondly, even though the book focuses on econometrics, I neglected some approaches that might warrant discussion. For example, there is hardly discussion of non-parametric inference, bootstraps, or even many specific cross-section topics. Many of the fundamental themes discussed here apply equally to econometric approaches which are beyond the scope of this book.

Thirdly, I have assumed that the readers of this book will be econometricians who are interested in the philosophical and statistical roots of their activities, and economic methodologists who have an interest in the scope and limits of empirical econometrics. This brings the risk that econometricians may find the econometric discussions not always satisfactory, while methodologists might complain that I have not done justice to all philosophical theories and subtleties that they can think of. I hope that both types of readers are willing to search for added value rather than concentrate on what they already know.

After the disclaimers finally some words of thanks. The book grew out of a PhD thesis, which was defended at the CentER for Economic Research of Tilburg University (The Netherlands). I also was able to discuss various parts with colleagues during visits to the London School of Economics, Duke University, the Eidgenössische Technische Hochschule Zürich, the University of Western Australia and at many seminars and conferences. For their financial support, I would like to thank in particular the Foreign and Commonwealth Office, the Fulbright Commission and NWO (Netherlands Organisation for Scientific Research).

In addition, I would like to thank many persons for inspiration, comments and collaboration. They are my fellow authors Ton Barten, Jan Magnus and Michael McAleer; many sparring partners among whom Mark Blaug, Nancy Cartwright, James Davidson, Tony Edwards, David Hendry, Colin Howson, Rudolf Kalman, the late Joop Klant, Ed Leamer, Neil de Marchi, Mary Morgan, Stephen Pollock, Ron Smith, Mark Steel, Paul Vitányi and Arnold Zellner. Michiel de Nooij provided highly appreciated research assistance.

1 The philosophy of induction

> [S]ome other scientists are liable to say that a hypothesis is definitely proved by observation, which is certainly a logical fallacy; most statisticians appear to regard observations as a basis for possibly rejecting hypotheses, but in no case for supporting them. The latter attitude, if adopted consistently, would reduce all inductive inference to guesswork.
>
> Harold Jeffreys ([1939] 1961, p. ix)

1 Introduction

Occasionally, the aspirations of econometrics are frustrated by technical difficulties which lead to increasing technical sophistication. More often, however, deeper problems hamper econometrics. These are the problems of scientific inference – the logical, cognitive and empirical limitations to induction. There is an escapist tendency in econometrics, which is to seek salvation in higher technical sophistication and to avoid deeper philosophical problems. This is reflected by the erosion of an early foothold of empirical econometrics, *Econometrica*. The share of empirical papers has declined from a third in the first (1933) volume to a fifth in recent volumes. This is not because most empirical values for economic variables or parameters have been settled. Despite the 'econometric revolution', there is no well established numerical value for the price elasticity of bananas. If *Econometrica* were to publish an issue with well established econometric facts, it might be very thin indeed. The factual knowledge of the economy remains far from perfect, as are the ability to predict its performance, and the understanding of its underlying processes. Basic economic phenomena, such as the consumption and saving patterns of agents, remain enigmatic. After many years of econometric investigation, there is no agreement on whether money causes output or not. Rival theories flourish. Hence, one may wonder what the added-value of econometrics is. Can we learn from experience in economics, and, if so, does econometrics itself serve this purpose? Or, were the aspirations too high after all, and does the sceptical attitude of Keynes half a century ago remain justified today?

2 Humean scepticism

An important issue in the philosophy of science is how (empirical) knowledge can be obtained.[1] This issue has a long history, dating back (at least) to the days of the Greek Academy, in particular to the philosopher Pyrrho of Elis (*c.* 365–275 BC), the first and most radical sceptic. Academic scepticism, represented for example by Cicero (106–43 BC), is more moderate than Pyrrho's. The ideas of Pyrrho (who did not write books, 'wisely' as Russell, 1946, p. 256, remarks) are known via his pupil Timon of Phlius (*c.* 320–230 BC) and his follower Sextus Empiricus (second century AD), whose work was translated into Latin in 1569. A few earlier translations are known but they have probably only been read by their translators. The 1569 translation was widely studied in the sixteenth and seventeenth centuries. All major philosophers of this period referred to scepticism. René Descartes, for example, claimed to be the first philosopher to refute scepticism.

One of the themes of the early sceptics is that only deductive inference is valid (by which they mean: logically acceptable) for a demonstrative proof, while induction is invalid as a means for obtaining knowledge. Perception does not lead to general knowledge. According to Russell (1946, p. 257),

> Scepticism naturally made an appeal to many unphilosophic minds. People observed the diversity of schools and the acerbity of their disputes, and decided that all alike were pretending to knowledge which was in fact unattainable. Scepticism was a lazy man's consolation, since it showed the ignorant to be as wise as the reputed men of learning.

Still, there was much interest in scepticism since the publication of the translation of Sextus Empiricus' work, not only by 'unphilosophic minds'. Scepticism has been hard to refute. Hume contributed to the sceptical doctrine (although he did not end up as a Pyrrhonian, *i.e.* radical sceptic). The result, 'Humean scepticism', is so powerful, that many philosophers still consider it to be a death blow to induction, the 'scandal of philosophy'.[2]

Hume ([1739] 1962) argues that the empirical sciences cannot deliver causal knowledge. There are no rational grounds for understanding the causes of events. One may observe a sequence of events and call them cause and effect, but the connection between the two remains hidden. Generalizations deserve scepticism. Hume (Book I, Part III, section 12, p. 189) summarizes this in two principles:

> *that there is nothing in any object, considered in itself, which can afford us a reason for drawing a conclusion beyond it;* and, *that even after the observation of the*

frequent or constant conjunction of objects, we have no reason to draw any inference concerning any object beyond those of which we have had experience.

The 'scandal of philosophy' is fundamental to empirical scientific inference (not just econometrics). It has wider implications (as Hume indicates) than denying causal inference. For example, does past experience justify the expectation of a sunrise tomorrow? The question was raised in discussing the merits of Pierre Simon de Laplace's 'rule of succession', a statistical device for induction (see chapter 2).[3] Another example, popular in philosophy, deals with extrapolation to a population instead of the future: if only white swans have been observed, may we infer that all swans are white? (This is the classic example of an affirmative universal statement.)

The sceptical answer to these questions is negative. The rules of deductive logic prohibit drawing a general conclusion if this conclusion is not entailed by its propositions. There is no *logical* reason why the next swan should be white. Of course, swans can be defined to be white (like statisticians who define a fair die to be unbiased), making black swans a contradiction in terms. An alternative strategy is to conclude that all *known* swans are white. The conclusion is conditional on the observed sample. Hence, the choice is between formulating definitions or making conditional enumerations. But most empirical scientists want to make generalizations. This is impossible if the induction problem proves insurmountable. Therefore, an understanding of induction is essential.

The logical form of the induction problem is that all *observed* X are Φ does not entail that all X are Φ. The next three chapters, dealing with probabilistic inference, consider a more delicate, probabilistic form of the induction problem: given that *most* observed X are Φ, what can be said about X in general? The source of Humean scepticism follows from the conjunction of three propositions (Watkins, 1984, p. 3):

(i) there are no synthetic *a priori* truths about the external world;
(ii) any genuine knowledge we have of the external world must ultimately be derived from perceptual experience;
(iii) only deductive derivations are valid.

The conjunction of (i), (ii) and (iii) does not allow for inferring knowledge beyond the initial premises. In this sense, inductive inference is impossible.

John Watkins (p. 12) argues that a philosophical, or 'rational' answer to scepticism is needed, because otherwise it is likely to encourage irrationality. Watkins holds that Hume himself regarded philosophical scepticism as an academic joke. Indeed, Hume uses the expression *jeux d'esprit* (in *A letter From a Gentleman to his Friend in Edinburgh*, included as an appendix to Hume [1748] 1977, p. 116). Describing the person who is afflicted by Pyrrhonism, Hume (p. 111) concludes:

> When he awakes from his dream, he will be the first to join in the laugh against himself, and to confess, that all his objections are mere amusement.

Amusement, Watkins (1984, p. 12) argues, does not qualify as a rational answer to scepticism. In fact, Hume's response is more elaborate than the quotation suggests. It relies on conventionalism (see below). I agree with Watkins that, formally, conventionalism is not very appealing (although conventions have much practical merit). Fortunately, there are alternatives. Once the source of Hume's problem (the threefold conjunction just mentioned) is clarified, the merits of those alternative responses to scepticism can be appraised.

Watkins (pp. 4–5) discusses a number of strategies as responses to Hume's problem. The most interesting ones are:

- the naturalist (ignoring the conjunction of propositions (i)–(iii));
- the apriorist (denying proposition (i));
- the conjecturalist (amending proposition (ii)); and
- the probabilist strategy (which takes odds with proposition (iii)).

A more detailed discussion of the probabilist strategy will be given in the next three chapters, while the remainder of this book considers how well this strategy may work in econometrics.

3 Naturalism and pragmatism

Descartes argued that one should distrust sensations. Insight in causal relations results from mere reasoning. Hume, criticizing Cartesian 'dogmatic rationalism', argues that such plain reasoning does not suffice to obtain unique answers to scientific questions. Cartesian doubt, 'were it ever possible to be attained by any human creature (as it plainly is not) would be entirely incurable' (Hume [1748] 1977, p. 103). It would not yield true knowledge either: 'reasoning *a priori*, any thing might appear able to produce anything' (*Letter From a Gentleman*, p. 119). Cartesian doubt is unacceptable to Hume ([1739] 1962, p. 318). It gave him a headache:

> The *intense* view of these manifold contradictions and imperfections in human reason has so wrought upon me, and heated my brain, that I am ready to reject all belief and reasoning, and can look upon no opinion even as more probable or likely than another.

But this does not make Hume a Pyrrhonian or radical sceptic. He is rescued from this philosophical 'melancholy and delirium' by nature.

His naturalist strategy is to concede that there is no epistemological answer to scepticism, but to deny its importance. It is human nature to make generalizing inferences, the fact that inference is not warranted from a logical point of view has no practical implications. Hume (*An Abstract of a Book Lately Published, Entitled, A Treatise of Human Nature, Etc.*, in Hume [1739] 1962, p. 348) concludes,

> that we assent to our faculties, and employ our reason only because we cannot help it. Philosophy would render us entirely *Pyrrhonian*, were not nature too strong for it.

The great subverter of Pyrrhonism, Hume ([1748] 1977, p. 109) writes, is 'action, and employment, and the occupations of common life'. Not reasoning, but custom and habit, based on the awareness of constant conjunctions of objects, make human beings draw inferences (p. 28). This response is known as conventionalism. According to Hume (p. 29), custom is the 'great guide of human life', and without custom or habit, those who are guided only by Pyrrhonian doubt will 'remain in a total lethargy, till the necessities of nature, unsatisfied, put an end to their miserable existence' (p. 110). Reason is the slave of our passions.

A pinch of Pyrrhonian doubt remains useful, because it makes investigators aware of their fallibility (p. 112). The fact that one cannot obtain absolute certainty by human reasoning does not imply universal doubt, but only suggests that researchers should be modest (*Letter From a Gentleman*, p. 116). But many scientists will feel embarrassed by the conclusion that custom is the ultimate foundation of scientific inference. Watkins, for example, rejects it. However, conventionalism may be rationally justified. This has been attempted by some adherents of the probabilistic approach. Other strategies related to Hume's conventionalism are instrumentalism (developed by John Dewey) and pragmatism, or pragmaticism, as Charles Peirce christened it. These hold that hypotheses may be accepted and rejected on rational grounds, on the basis of utility or effectiveness. The pragmatic approach can be combined with the probabilistic strategy. But it is not free of problems. Most importantly, it is an invitation to scientific obscurantism (should a theory be useful to the learned – who qualifies? – or to the mighty?). A problem with conventionalism is to give an answer to the question 'where do these conventions come from?' and to provide a rational justification for the conventions (evolutionary game theory has been directed to this question). Lawrence Boland (1982; also 1989, p. 33) argues that neoclassical economists deal with the induction problem by adopting a conventionalist strategy. Econometricians base much of their work on another convention concerning the size of a test: the well known 5% significance level. This convention has its roots in a quarrel between Karl Pearson and R. A. Fisher, two founders of modern statistics (see chapter 3, section 3.2).

4 Apriorism

The *apriorist strategy* to the problem of scepticism denies proposition (i), concerning the absence of synthetic *a priori* truth. Immanuel Kant invented this notion of *a priori* synthetic truth, true knowledge that is both empirical and based on reasoning. It is neither analytic nor synthetic.[4] The canonical example of an *a priori* synthetic truth is Kant's Principle of Universal Causation, which is his response to Humean scepticism. Kant argued that everything must have a cause: 'Everything that happens presupposes something upon which it follows in accordance with a rule' (translated from *Kritik der reinen Vernunft*, Kant's most important work, published in 1781; in Krüger, 1987, p. 72). This doctrine is also known as causal determinism, or simply as causalism (Bunge [1959] 1979, p. 4).

Unlike Hume, John Stuart Mill endorsed Kant's principle: for Mill, induction is the search for causes. Mill distinguishes four 'canons of induction', given in his *Logic*, III (viii); Mill [1843] 1952):

- the method of agreement;
- the method of difference;
- the method of residues;
- the method of concomitant variations.

These methods are based on the 'principle of uniformity of nature', which holds that the future will resemble the past: the same events will happen again if the conditions are sufficiently similar. The method of difference starts from the premise that all events have a cause. The next step is to give an exhaustive list of possible causes, and select the one(s) which always occurs in common with the event, and does not occur if the event does not occur. A problem is to select this exhaustive list of possible causes.

Keynes ([1921] CW VIII, p. 252) refers to the principle of uniformity of nature in his discussion of reasoning by analogy, and suggests that differences in position in time and space should be irrelevant for the validity of inductions. If this principle forms the basis for induction, it cannot itself be founded upon inductive arguments. Furthermore, it is doubtful that experience validates such a strong principle. Nature seems much more erratic and surprising than the principle of uniformity of nature suggests. Still, the late philosopher Karl Popper ([1935] 1968, p. 252) explicitly argues that 'scientific method presupposes *the immutability of natural processes*, or the "principle of the uniformity of nature"'.

Likewise, Bernt Stigum (1990, p. 542) argues that this principle is a necessary postulate of epistemology. Some probability theorists advocate a statistical version of this synthetic *a priori* truth: the stability of mass phenomena (see in particular the discussion of von Mises in chapter 3, section 2).

In the social sciences, it is not the uniformity of nature which is of interest, but the relative stability of human behaviour. A more apt terminology for the principle would then be the 'principle of stable behaviour'. Consider the axioms of consumer behaviour. If one assumes that preferences are stable (Hahn, 1985, argues this is all the axioms really say), then accepting these axioms as *a priori* truths warrants inductive generalizations. This principle solves, or rather, sidesteps, the Humean problem. If it is accepted, generalizations from human experience are admissible. But again this postulate is doubtful. Too frequently, humans behave erratically, and on a deeper level, reflexivity (self-fulfilling prophecies) may undermine uniform regularities in the social sciences. It suffers from the same problems as the principle of uniformity of nature: either it is false, or its justification involves infinite regress. But a weaker principle of stable behaviour may be accepted, by giving a probabilistic interpretation to the generalization. There should be an appreciable (non-zero) probability that stable behaviour may be expected. This is the basis for rational behaviour. A fair amount of stability is also necessary (not sufficient) for scientific inference: otherwise, it is impossible to 'discover' laws, or regularities.

It is hard to imagine interesting *a priori* synthetic truths specific to economics. The axioms of consumer behaviour are not generally accepted as true. An investigation of their validity cannot start by casting them beyond doubt (chapter 8 provides a case history of 'testing' consumer demand theory). Bruce Caldwell (1982, p. 121) discusses praxeological axioms of Austrian economists as an example of Kant's *a priori* synthetical propositions. The Austrian Friedrich von Wieser argued that a cumbersome sequence of induction is not needed to establish laws in economics. He claimed (cited in Hutchison, 1981, p. 206) that we can 'hear the law pronounced by an unmistakable inner voice'. Ludwig von Mises made apriorism the cornerstone of his methodology. The problem of this line of thought is that inner voices may conflict. If so, how are we to decide which voice to listen to?

5 Conjecturalism

The *conjecturalist strategy* denies Watkins' proposition (ii) and instead holds that scientific knowledge is only negatively controlled by experience: through falsification. Popper provided the basic insights of the conjecturalist philosophy (also known as methodological falsificationism) in his *Logik der Forschung* in 1934 (translated as Popper, [1935] 1968). This nearly coincides with one of the first efforts to test economic theory with econometric means (Tinbergen, 1939b). Followers of Popper are, among others, Imre Lakatos and Watkins. I will first discuss Popper's views on inference, then Lakatos' modified conjecturalism.

5.1 Popper's conjecturalism

Popper's impact on economic methodology has been strong. Two pronounced Popperians in economics are Mark Blaug (1980) and Terence Hutchison (1981). Moreover, statisticians and econometricians frequently make favourable references to Popper (Box, 1980, p. 383, n.; Hendry, 1980; Spanos, 1986) or believe that Popper's is 'the widely accepted methodological philosophy as to the nature of scientific progress' (Bowden, 1989, p. 3). Critics claim that the real impact of Popperian thought on economic inference is more limited (see also De Marchi, 1988; Caldwell, 1991).

5.1.1 Falsification and verification

Scientific statements are those which can be refuted by empirical observation. Scientists should make bold conjectures and try to falsify them. This is the conjecturalist view in a nutshell. More precisely, theories are thought of as mere guesses, conjectures, which have to be falsifiable in order to earn the predicate scientific. The *modus tollens* (if *p*, then *q*. But not-*q*. Therefore, not-*p*) applies to scientific inference – if a prediction which can be deduced from a generalization (theory) is falsified, then that generalization itself is false. The rules of deductive logic provide a basis for scientific rationality and, therefore, make it possible to overcome the problems of Humean scepticism. Falsifiability distinguishes science from non-science (the demarcation criterion). The growth of knowledge follows from an enduring sequence of conjectures and refutations. Theories are replaced by better, but still fallible, theories. Scientists should remain critical of their work.

So far, there seems not much controversial about the conjecturalist approach. The tentative nature of science is a commonplace. Popper went beyond the commonplace by constructing a philosophy of science on it, methodological falsificationism. A source of controversy is Popper's critique of logical positivism, the philosophy associated with the *Wiener Kreis*.[5] A related source is his obnoxious rejection of induction.

Logical positivism holds that the possibility of empirical *verification*, rather than falsification, makes an empirical statement 'meaningful' (the meaning lies in its method of verification). There are many problems with this view, but Popper aimed his fire at an elementary one: affirmative universal statements, like 'all swans are white', are not verifiable. In response to Popper's critique, Carnap dropped the verifiability criterion and started to work on a theory of confirmation (see also chapter 4, section 3.1). Again, this theory was criticized by Popper.

The logical difference between verification and falsification is straightforward. The observation of a white swan does not imply the truth of the claim 'all swans are white'. On the other hand, observing a black swan makes a judgement about the truth of the claim possible. In other words, there is a logical asymmetry between verification and falsification. This asymmetry is central to Popper's ideas: 'It is of great importance to current discussion to notice that falsifiability in the sense of my demarcation principle is *a purely logical affair*' (Popper, 1983, p. xx; emphasis added). This logical affair is not helpful in guiding the work of applied scientists, like econometricians. It should have real-world implications. For this purpose, Popper suggests the crucial test, a test that leads to the unequivocal rejection of a theory. Such a test is hard to find in economics.

According to Popper, it is much easier to find confirmations than falsifications. In the example of swans this may be true, but for economic theories things seem to be rather different. It is not easy to construct an interesting economic theory which cannot be rejected out of hand. But if verification does not make science, Popper needs another argument for understanding the growth of knowledge. Popper ([1935] 1968, p. 39) bases this argument on severe testing:

> there is a great number – presumably an infinite number – of 'logically possible worlds'. Yet the system called 'empirical science' is intended to represent only *one* world: the 'real world' or 'world of our experience'... But how is the system that represents our world of experience to be distinguished? The answer is: by the fact that it has been submitted to tests, and has stood up to tests.

Experience is the sieve for the abundance of logically possible worlds. The difference with induction results from a linkage of experience with falsifications: experience performs a *negative* function in inference, not the positive one of induction.

Popper's idea that the truth of a theory cannot be proven on the basis of (affirming) observations, is not revolutionary – indeed, it basically rephrases Hume's argument. Obviously, it was known to the logical positivist. And it had already been a common-sense notion in the statistical literature for ages (in fact, Francis Bacon had already made the argument, as shown by Turner, 1986, p. 10). One can find this, explicitly, in the writings of Karl Pearson, Ronald Aylmer Fisher, Harold Jeffreys (see the epigraph to this chapter), Jerzy Neyman and Egon Pearson,[6] Frederick Mills, Jan Tinbergen,[7] Tjalling Koopmans (1937) and probably many others. They did not need philosophical consultation to gain this insight, neither did they render it a philosophical dogma according to which falsification becomes the highest virtue of a scientist.[8] Econometricians are, in this respect, just like other scientists: they rarely aim at falsifying, but try to construct satisfactory empirical models (see Keuzenkamp and Barten, 1995). Of course, 'satisfactory' needs to be defined, and this is difficult.

Jeffreys' epigraph to this chapter can be supplemented by a remark made by the theoretical physicist Richard Feynman (1965, p. 160): 'guessing is a dumb man's job'. A machine fabricating random guesses may be constructed, consequences can be computed and compared with observations. Real science is very different: guesses are informed, sometimes resulting from theoretical paradoxes, sometimes from experience and experiment. Jeffreys argues that one may agree with Popper's insight that confirmation is not the same as proof, without having to conclude that confirmation (or verification) is useless for theory appraisal, and induction impossible.

5.1.2 The crucial test

An important example to illustrate Popper's ([1935] 1968) methodological falsificationism is Einstein's general theory of relativity, which predicts a red shift in the spectra of stars.[9] This is the typical example of a prediction of a novel fact which can be tested. Indeed, a test was performed with a favourable result. But Paul Feyerabend (1975, p. 57, n. 9) shows that Einstein would not have changed his mind if the test had been negative. In fact, many of Popper's examples of crucial tests in physics turn out to be far more complicated when studied in detail (see Feyerabend, 1975; Lakatos, 1970; Hacking, 1983, chapter 15, agrees with Lakatos' critique on crucial tests, but criticizes Lakatos for not giving proper credit to empirical work).

For several reasons, few tests are crucial. First, there is the famous 'Duhem–Quine problem'. Second, in many cases rejection by a 'crucial test' leaves the researcher empty handed. It is, therefore, unclear what the implication of such a test should be – if any. Third, most empirical tests are probabilistic. This makes it hard to obtain decisive inferences (this will be discussed below).

The Duhem–Quine problem is that a falsification can be a falsification of anything. A theory is an interconnected web of propositions. Quine ([1953] 1961, p. 43) argues,

> Any statement can be held true come what may, if we make drastic enough adjustments elsewhere in the system.... Conversely, by the same token, no statement is immune to revision. Revision even of the logical law of the excluded middle has been proposed as a means of simplifying quantum mechanics; and what difference is there in principle between such a shift and the shift whereby Kepler superseded Ptolemy, or Einstein Newton, or Darwin Aristotle?

For example, rejection of homogeneity in consumer demand (see chapter 8) may cast doubt on the homogeneity proposition, but also point to problems due to aggregation, dynamic adjustment, the quality of the data and so on. In the example 'all swans are white' the observation of a green swan may be a falsification, but also evidence of hallucination or proof that the observer wears sunglasses. The theory of simplicity provides useful additional insights in the Duhem–Quine thesis (see chapter 5, section 3.2).

Second, what might a falsification imply? Should the theory be abandoned? Or, if two conflicting theories are tested, how should falsifications be weighted if both theories have defects? This is of particular interest in economics, as no economic theory is without anomalies. If induction is impossible, is the support for a theory irrelevant? Popper ([1963] 1989, chapter 10) tries to formulate an answer to these questions by means of the notion of verisimilitude or 'truthlikeness'. In order to have empirical content, verisimilitude should be measurable. But this brings in induction through the back door (see chapter 3, section 5.3).

In a review of Popper's methodology, Hausman (1988, p. 17) discusses the first and second problem. He concludes that Popper's philosophy of science is 'a mess, and that Popper is a very poor authority for economists interested in the philosophy of science to look to'. Hausman shows that a Popperian either has to insist on logical falsifiability, in which case there will be no science (everything will be rejected), or has to consider entire 'test systems' (in Popper's vocabulary, 'scientific systems'), in which case severe testing has little impact on the hypothesis of interest. The reason for the latter is that such a test system combines a number of basic statements and auxiliary hypotheses. If, as Popper claims, confirmation is impossible, one is unable to rely on supporting evidence from which one may infer that the auxiliary hypotheses are valid. A falsification, therefore, can be the falsification of anything in the test system: the Duhem–Quine problem strikes with full force. One should be able to rely on some rational reason for tentative acceptance of 'background knowledge'. Crucial tests and logical falsification are of little interest in economic inference. The complications of empirical testing, or what Popper also calls 'conventional falsification', are much more interesting, but Popper is of little help in this regard. Lakatos (1978, pp. 165–6) reaches a similar conclusion as Hausman:

> By refusing to accept a 'thin' metaphysical principle of induction Popper fails to separate rationalism from irrationalism, weak light from total darkness. Without this principle Popper's 'corroborations' or 'refutations' and my 'progress' or 'degeneration' would remain mere honorific titles awarded in a pure game... only a positive solution of the problem of induction can save Popperian rationalism from Feyerabend's epistemological anarchism.

Lakatos' own contribution is evaluated in section 5.2.

The third problem of crucial tests, the probabilistic nature of empirical science, is of particular interest in econometrics. Popper ([1935] 1968, p. 191) notes that probability propositions ('probability estimates' in his words) are not falsifiable. Indeed, Popper (p. 146) is aware that this is an

> almost insuperable objection to my methodological views. For although probability statements play such a vitally important role in empirical science, they turn out to be in principle *impervious to strict falsification.* Yet this stumbling block will become a touchstone upon which to test my theory, in order to find out what it is worth.

Popper's probability theory is discussed in chapter 3, section 5. There, I argue that it is unsatisfactory, hence the stumbling block is a painful one. One objection can already be given. In order to save methodological falsificationism, Popper proposes a methodological rule or convention for practical falsification: to regard highly improbable events as ruled out or prohibited (p. 191; see Watkins, 1984, p. 244 for support, and Howson and Urbach, 1989, p. 122 for a critique). This is known as *Cournot's rule*, after the mathematician and economist Antoine-Augustin Cournot, who considered highly improbable events as physically impossible. Cournot's rule has been defended by probability theorists such as Emile Borel and Harald Cramér (1955, p. 156), without providing a deeper justification. The rule has even been used to support the hypothesis of divine providence: life on earth is highly improbable and must be ruled out, if not for the existence of the hand of God.

The problem with the rule is where to draw the line: when does improbable turn into impossible? Popper ([1935] 1968, p. 204) argues that a methodological rule might decree that only reasonably fair samples are permitted, and that predictable or reproducible (*i.e.* systematic) deviations must be ruled out. But the concept of a fair sample begs the question. If an experiment can be repeated easily (in economics, controlled experiments are rare and reproducible controlled experiments even more so), this may be a relatively minor problem, but otherwise it can lead to insoluble debates. Is a test result, with an outcome that is improbable given the hypothesis under consideration, a fluke or a straightforward rejection? If ten coins are tossed 2,048 times, each particular sequence is extremely improbable, and, by Cournot's principle, should be considered impossible.[10] Cournot's rule does not provide sound guidance for the following question: What will be a falsification of the statement that most swans are white, or 'Giffen goods are rare'? This type of proposition will be discussed in the treatment of the probabilistic strategy (their empirical content, not their falsifiability, is of practical interest).

5.1.3 Critical rationalism

Caldwell (1991, p. 28) argues that, despite its drawbacks, Popper's falsificationism partly captures the spirit of economic inference. The fact, however, that economists occasionally test theories and sometimes conclude that they reject something, does not imply that methodological falsificationism is an apt description of this part of economic inference. Methodological falsificationism deals with a logical criterion to be applied to logical caricatures of scientific theories (in this sense, Popper operates in the tradition of the *Wiener Kreis*). There are neither crucial tests in economics, nor is there a strong desire for falsifications.

Caldwell continues that falsificationism may be abandoned, but that Popper's true contribution to scientific method is to be found in critical rationalism. This is a much weakened version of methodological falsificationism and merely implies that scientists should be (self-) critical. It reduces Popper's philosophy to a platitude (some authors indeed claim that Popper trivializes science, e.g. Feyerabend, 1987). Unlike methodological falsificationism, critical rationalism purports to be a historically accurate description of science. By showing that, in a number of historical cases, good science accords to critical rationalism, Popper suggests that the stronger programme of methodological falsificationism is supported. But those case studies provide weak evidence (Hacking, 1983), or even are 'myths, distortions, slanders and historical fairy tales' (Feyerabend, 1987, p. 185). This does not mean that critical rationalism is invalid; on the contrary, it is a principle which has been advocated by a wide range of scientists, before and after the publication of Popper's views on methodology.

For example, the motto of Karl Pearson ([1892] 1911) is a statement due to Victor Cousin: '*La critique est la vie de la science*' (*i.e.* criticism is the life of science; see also Pearson, p. 31, among many other places where he emphasizes the importance of criticism). More or less simultaneously with Pearson, Peirce introduced the theory of fallibilism, of which falsificationism is a special case (see e.g. Popper, [1963] 1989, p. 228). According to fallibilism, research is stimulated by a state of unease concerning current knowledge. Research intends to remove this state of unease by finding answers to scientific puzzles. In Humean vein, Peirce (1955, p. 356) argues, 'our knowledge is never absolute but always swims, as it were, in a continuum of uncertainty and of indeterminacy'. Methodological falsificationism is different from Peirce's philosophy, by actually *longing* for a state of unease rather than attempting to remove it.

5.1.4 Historicism

Although Popper's philosophy of science is in many respects problematic, in particular if applied to the social sciences, he has contributed an important insight to the social sciences that is usually neglected in discussions of (economic) methodology. This is his critique of 'historicism'. Historicism, Popper ([1957] 1961) argues, starts from the idea that there is such a thing as a historical necessity of events, or social predestination. He considers Marx and Hegel as examples of historicists (although they had a very different interpretation of historicism and Popper's interpretation of their work is controversial).[11] Disregarding the historicist merits of Marx and Hegel, this may be Popper's most interesting book for economists.

Popper's critique of historicism is related to his idea that knowledge is conjectural and universal theories cannot be proven. This does not preclude growth of knowledge, but this growth results from trial and error. It cannot be predicted. The ability to predict the future course of society is limited as it depends on the growth of knowledge.

One of the most interesting parts in the *Poverty of Historicism* is his discussion of the so-called *Oedipus Effect*. Popper (p. 13; see also Popper, 1948) defines it as:

> the influence of the prediction upon the predicted event (or, more generally, . . . the influence of an item of information upon the situation to which the information refers), whether this influence tends to bring about the predicted event, or whether it tends to prevent it.

This problem of reflexivity, as it is also known, undermines methodological falsificationism. Popper's views on historicism and the special characteristics of the social sciences suggest that falsificationism cannot play quite the same role in the social sciences as in the natural sciences. Popper

([1957] 1961, pp. 130–43) is ambiguous on this point. He remains convinced of the general importance of his methodology, the unity of method. Meanwhile, he recognizes the difference between physics and economics:

> In physics, for example the parameters of our equations can, in principle, be reduced to a small number of natural constants – a reduction that has been carried out in many important cases. This is not so in economics; here the parameters are themselves in the most important cases quickly changing variables. This clearly reduces the significance, interpretability, and testability of our measurements. (p. 143)

Popper swings between the strong imperatives of his own methodology, and the more reserved opinions of his former colleagues at the London School of Economics (LSE), Lionel Robbins and Friedrich Hayek. Popper's compromise consists of restricting the domain of the social sciences to an inquiry for conditional trends or even singular events, given an *a priori* axiom of full rationality.[12] Popper calls this the zero method. Methodological falsificationism may be applied to this construct, but the rationality postulate is exempted from critical scrutiny. It is an *a priori* synthetic truth (see also Caldwell, 1991, pp. 19–21).

The initial conditions may change and this may invalidate the continuation of the trend. Hence, social scientists should be particularly interested in an analysis of initial conditions (or situational logic). The difference between prognosis and prophecy is that prophecies are unconditional, as opposed to conditional scientific predictions (Popper, [1957] 1961, p. 128). In the social sciences, conditions may change due to the unintended consequence of human behaviour (this is a cornerstone of Austrian economic thought). Mechanic induction like extrapolation of trends is, therefore, not a very reliable way of making forecasts. The econometric implications of the Oedipus Effect and the lack of natural constants in economics deserve more attention. These implications are of greater interest to the economist or econometrician than the methodology of falsificationism or Popper's ideas on probability.

5.2 Lakatos and conjecturalism

What does Lakatos offer to rescue the conjecturalist strategy? Because of the problems involved with methodological falsificationism, he proposes to give falsifications less impact. He rejects the crucial test or 'instant falsification', and instead emphasizes the dynamics of theory development.

5.2.1 Research programmes

This dynamics can be evaluated by considering a theory as a part of an ongoing *research programme*. A theory is just one instance of a research programme, RP, at a given point in time. How do you decide whether a succeeding theory still belongs to an RP? This question should be settled by defining the essential characteristics of an RP, by its *hard core* and the guidelines for research, the *heuristic*. The hard core consists of the indisputable elements of a research programme. The positive heuristic provides the guidelines along which research should proceed. The negative heuristic of a research programme forbids directing the *modus tollens* at the hard core (Lakatos, 1978, p. 48).

According to Lakatos (p. 48), the hard core of an RP is irrefutable by the methodological decision of its proponents. A falsification of the theory is not automatically a rejection of an RP. Falsifying a theory is replaced by measuring the degree of progressiveness of an RP. Lakatos (p. 33) distinguishes three kinds of progress. A research programme is

- *theoretically* progressive if 'each new theory has excess empirical content over its predecessor, that is, if it predicts some novel, hitherto unexpected fact'
- *empirically* progressive if some of these predictions are confirmed
- *heuristically* progressive if it avoids auxiliary hypotheses that are not in the spirit of the heuristic of a research programme (Lakatos would call such hypotheses *ad hoc*$_3$, where *ad hoc*$_1$ and *ad hoc*$_2$ denote lack of theoretical and empirical progressiveness, respectively).

It is not easy in ecomonics to apply Lakatos' suggestion of appraising theories by comparing their rate of progressiveness or degeneration. A research programme is a vague notion. Scientists may disagree about what belongs to a specific *RP* and what does not (see Feyerabend, 1975). The problem culminates in the so-called tacking paradox (see Lakatos, 1978, p. 46). If the theory of diminishing marginal utility of money is replaced by a successor, which combines this theory with the general theory of relativity, an apparently progressive step is being made. Of course, this is not what Lakatos has in mind: this is why he emphasizes the consistency with the positive heuristic of an RP. An alternative for avoiding nonsensical combination of two theories is to introduce the notion of *irrelevant conjunction* (see Rosenkrantz, 1983, for a discussion and a Bayesian solution to the problem).

Lakatos' suggestions are of some help for understanding ('rationally reconstructing') economic inference, but in many cases they are too vague and insufficiently operational. Chapter 8 provides a case study showing how difficult it is to apply them to an important episode in the history of applied econometrics: testing homogeneity of consumer demand.

Competing research programmes may apply to partly non-overlapping areas of interest. This leads to problems such as the already mentioned Duhem–Quine problem, and *incommensurability,* Thomas Kuhn's (1962) notion that new theories yield new interpretations of events, and even of the language describing these events. Furthermore, whereas Popper still made an (unsuccessful) attempt to contribute to the theory of probabilistic inference (cf. his propensity theory of probability, and the notion of verisimilitude), Lakatos has a critical attitude towards this subject.

5.2.2 Growth and garbage

Both Popper and Lakatos attack inductivism, but Lakatos is more radical in his critique of empirical tests. These tests are a key element in Popper's epistemology, but not so in Lakatos'. This is clear from the *Methodology of Scientific Research Programmes*, in which falsifications are less crucial than in Popper's work. Instead of falsifiability and crucial tests, Lakatos advocates a requirement of growth. Can statistical methods help to obtain knowledge about the degree of empirical progress of economic theories or research programmes? Lakatos' scant remarks on this issue provide little hope (all quoted from Lakatos, 1970, p. 176). To start, the requirement of continuous growth

> hits patched-up, unimaginative series of pedestrian 'empirical' adjustments which are so frequent, for instance, in modern social psychology. Such adjustments may, with the help of so-called 'statistical techniques', make some 'novel' predictions and may even conjure up some irrelevant grains of truth in them. But this theorizing has no unifying idea, no heuristic power, no continuity.

These uncharitable statements are followed by such terms as *worthless*, *phoney corroborations*, and, finally, *pseudo-intellectual garbage*. Lakatos concludes:

> Thus the methodology of research programmes might help us in devising laws for stemming this intellectual pollution which may destroy our cultural environment even earlier than industrial and traffic pollution destroys our physical environment.

Whoever is looking for a Lakatosian theory of testing will have a hard job. Lakatos' approach is nearly anti-empirical. In his case studies of physics, experimenters are repeatedly 'taught lessons' by theoreticians; bold conjectures are made despite seemingly conflicting empirical evidence, and so on (see Hacking, 1983, chapter 15, for a devastating critique of Lakatos' account of many of these experiments).

Lakatos' approach is of little help for economists and, as I will argue later on (chapter 8), it does not provide a basis for a useful econometric methodology. If testing and falsifying of all propositions are approached rigorously, not much of economics (or any other science) will be left. On the other hand, if induction is rejected and severe testing as well, one ends up with 'anything goes' or 'anarchism in disguise' (Feyerabend, 1975, p. 200). Watkins (1984, p. 159), the saviour of methodological falsificationism, agrees with this characterization of Feyerabend and makes the following *reductio ad absurdum*:

> If you could tell, which you normally cannot, that Research Program 2 is doing better than Research Program 1, then you may reject Research Program 1, or, if you prefer, continue to accept Research Program 1.

What is missing in the conjecturalist strategy is a clear view on the utility of theories, the economy of scientific research (emphasized by Peirce) and the positive role of measurement and evidence. A theory of inductive reasoning remains indispensable for understanding science.

6 Probabilism

The probabilistic strategy may offer garbage, if applied badly (as Lakatos suggests) but it may also yield insights in scientific inference: in its foundations and in appraising econometric applications.

This brings us to the probabilist strategy, which claims that Hume wants too much if he requires a proof for the truth of inductive inferences, and amends proposition (iii) of section 2 above. A logic of 'partial entailment' is proposed: probability logic. This strategy has been investigated by John Maynard Keynes, Harold Jeffreys, Hans Reichenbach, Rudolf Carnap and many others. Alternatively, the problem of Humean scepticism may be resolved by providing a probabilistic underpinning of the principle of the uniformity of nature, which has been investigated with the Law of Large Numbers. This approach has been taken by Richard von Mises.

The following three chapters discuss these and other versions of a probabilistic strategy for scientific inference. The probabilistic strategy deserves special attention because it can serve as a foundation for econometric inference, the topic of this book. Furthermore, uncertainty is particularly important in economics. If falsifying economic theories is feasible at all, then it must be probabilistic falsification. The issue of probabilistic *testing* can only be well understood if a good account of probabilism is presented (note that testing is clearly neither the only nor the ultimate aim of probabilism). If probabilistic falsification is impossible, then other methods are needed for appraising economic theories given the uncertainty that is inextricably bound up with economics.

Notes

1. There are some excellent introductions to the philosophy of science, written for and by economists. Blaug (1980) and Caldwell (1982) are particularly strong on methodology, but both have little to say about econometrics. Darnell and Evans (1990) combine methodology and econometrics, but their treatment is brief and sometimes ambiguous. More recent developments in methodology can be found in Backhouse (1994).
2. The phrase is due to C. D. Broad (see Ramsey, 1926, p. 99; and Hacking, 1975, p. 31). Frank Ramsey (1926, pp. 98–9) denies that there is no answer to Hume's problem, but 'Hume showed that it [i.e. inductive inference] could not be reduced to deductive inference or justified by formal logic. So far as it goes his demonstration seems to be final; and the suggestion of Mr Keynes that it can be got round by regarding induction as a form of probable inference cannot in my view be maintained. But to suppose that the situation which results from this is a scandal to philosophy is, I think, a mistake.'
3. The problem is not trivial. Keynes ([1921] CW VIII, p. 418) notes how Laplace calculated that, 'account be taken of the experience of the human race, the probability of the sun's rising tomorrow is 1,826,214 to 1, this large number may seem in a kind of way to represent our state of mind of the matter. But an ingenious German, Professor Bobek, has pushed the argument a degree further, and proves by means of these same principles that the probability of the sun's rising every day for the next 4000 years, is not more, approximately, than two-thirds, — a result less dear to our natural prejudices.' See also Pearson ([1892] 1911, p. 141) for a discussion of Laplace and the probability of sunrise.
4. A proposition is analytic if the predicate is included in the subject (e.g. all econometricians are human). A synthetic proposition is not analytic (usually thought to be based on matters of fact; e.g. all econometricians are wise). Quine ([1953] 1961) contains a classic critique of the distinction.
5. The *Weiner Kreis* was the influential group of scientists who met regularly in Vienna during the 1920s and 1930s. In 1922 the group was formed by Moritz Schlick. Some prominent members were Rudolph Carnap, Otto Neurath and Hans Hahn. Other regular participants were Herbert Feigl, Philipp Frank, Kurt Gödel, Friedrich Waisman and Richard von Mises. The group built on the positivist doctrines of Henri Poincaré, and in particular the Austrian physicist and philosopher Ernst Mach. In addition, they used advances in logic due to Gottlob Frege, to Russell and Whitehead, whence logical positivism. A closely related view is logical empiricism, associated with Hans Reichenbach and, again, Carnap. Caldwell (1982) defines logical empiricism as the mature version of logical positivism. The different branches of twentieth-century positivism are often grouped as neo-positivism. Popper is occasionally associated with neo-positivism, but Hacking (1983, p. 43) argues convincingly that Popper does not qualify as a positivist.
6. Although with a twist in their case, as they are interested in behaviour rather than inference.

7. Tinbergen (1939a, p. 12) argues that econometrics cannot prove a theory right, but it may show that some theories are not supported by the data. He never refers to Popper, or any other philosopher (Popper, [1957] 1961, on the other hand, contains a reference to Tinbergen who notes that constructing a model is a matter of trial and error). Koopmans makes a remark similar to Tinbergen's. If researchers still speak of verification of theories, then this should not be taken literally: very few would deny Hume's argument.
8. According to Mark Blaug, an important difference between Popper and those statisticians is Popper's banning of immunizing stratagems. Whether an absolute ban would benefit science is doubtful. See Keuzenkamp and McAleer (1995) for a discussion of 'ad hocness' and inference, in particular the references to Jeffreys.
9. Popper regards Einstein as the best example of a scientist who took falsification seriously, because of his bold predictions. Einstein has the highest number of entries in the name index of Popper ([1935] 1968) (thirty-six times). Carnap (with a score of thirty-five) is a close second, but does not share in Popper's admiration.
10. The example is not imaginary. W. Stanley Jevons performed this experiment to 'test' Bernoulli's law of large numbers. This and more laborious though equally meaningless experiments are reported in Keynes ([1921] CW VIII, pp. 394–9).
11. Hegel, for example, believed in the historical relativity of truth, a tenable position from the point of view presented in this book.
12. '[A]nd perhaps also on the assumption of the possession of complete information' (Popper, [1957] 1961, p. 141) – a remarkably un-Austrian assumption!

2 Probability and indifference

> One regards two events as equally probable when one can see no reason that would make one more probable than the other.
>
> Laplace (cited in Hacking, 1975, p. 132)

1 Introduction

Are scientific theories 'probabilifiable'? Lakatos (1978, p. 20), who phrases this question, says no. According to the probabilistic response to Humean scepticism, however, the answer should be affirmative, although the justification for the affirmative answer is problematic. The crossroads of philosophy of science and probability theory will be the topic of chapters 2 to 4.

There exist different versions of probabilism. The roots of modern probability theory lie in Laplace's indifference theory of probability, which has been supported by economists such as John Stuart Mill and W. Stanley Jevons. Laplace's theory is complemented by the theory of the English Reverend Thomas Bayes. Their interpretation of probability, which is based on 'equally probable' events, will be presented in chapter 2, section 2.

Since the days of Bayes and Laplace, confusion has existed about whether probability refers to a property of events themselves, or beliefs about events. For this purpose, it is useful to make a distinction between the realist or aleatory interpretation of probability on the one hand, and the epistemological interpretation of probability on the other hand (Hacking, 1975).

Aleatory stems from the Latin word for 'die' (*alea jacta est*). The aleatory view has its roots in the study of games of chance, such as playing cards and casting dice. It relates probability to the occurrence of events or classes of events. The relative frequency theory of probability (in brief, frequency theory) is based on this view. The propensity theory, which regards probability as a physical property, also belongs to this interpretation. Both of these aleatory interpretations of probability are discussed in chapter 3. The frequency theory of probability dominates econometrics. An unfortunate misnomer, frequently used in this context, is 'classical econometrics (statistics)'. Classical statistics is the proper label for the indifference interpretation.

The epistemological view, which underlies Bayesian econometrics, is discussed in chapter 4. This view is concerned with the credibility of propositions, assertions and beliefs in the light of judgement or evidence (Hacking, 1975, p. 14). Within the epistemological view, there exist 'objective' (or logical) and 'subjective' (or personalistic) interpretations. The theory of Keynes is an example of a logical probability theory, while Frank Ramsey, Bruno de Finetti and Leonard ('Jimmie') Savage developed subjective theories of probability. Subjective theories of probability enable dealing with cognitive limitations to human knowledge and inference.

Although most of modern probability theory is based on a formalism due to A. N. Kolmogorov (presented in the Intermezzo), applied probability theory (statistical theory) cannot do without an *interpretation* of this formalism. Chapters 2 to 4 deal with this interpretation. It is an essential prerequisite for a proper understanding of different schools in econometrics and for an appreciation of methodological controversies among econometricians.

2 The indifference theory of probability

2.1 Indifference

The indifference or classical theory of probability dates back to Christiaan Huygens, Gottfried von Leibniz, Jacques Bernoulli and, especially, Pierre Simon de Laplace (see Hacking, 1975, pp. 126–127; von Mises, [1928] 1981, p. 67). This theory defines probability as the ratio of the number of favourable outcomes to the total number of equally likely outcomes.

More precisely, if a random experiment can result in n mutually exclusive and equally likely outcomes and if n_A of these outcomes have an attribute A, then the probability of A is

$$P(A) = \frac{n_A}{n}. \tag{1}$$

Note that this interpretation, if taken literally, restricts probabilities to rational numbers (the same applies to the frequency interpretation). This is not quite satisfactory, because it excludes the appraisal of two single independent and equally likely events if their joint probability is 1/2 (the single probability of each event, $1/\sqrt{2}$ has neither an indifference nor a frequency interpretation).

The indifference theory of probability has more serious drawbacks, and, therefore, has little support today. Even so, a discussion is useful for two reasons. First, the problems of the indifference theory stimulated the formulation of other theories of probability. Second, the indifference theory is relevant for understanding the problem of non-informative priors in Bayesian statistics.

The name 'principle of indifference' is due to Keynes ([1921] CW VIII), who preferred this label to 'the principle of insufficient reason' by which it was known before (see Hacking, 1975, p. 126). Jacques Bernoulli introduced the principle, which states that events are equally probable if there is no known reason why one event (or alternative) should occur (or be more credible) rather than another (Keynes, [1921] CW VIII, p. 44). This formulation allows an epistemological as well as an aleatory or realist interpretation. I will continue with the latter, but all drawbacks which will be discussed here apply to the epistemological one as well.

2.2 Objections

A first objection to the indifference interpretation of probability is that the definition is either circular, which reduces the definition of probability to a tautology, or incomplete. The circularity arises if 'equally likely' is identified with 'equally probable'. The probability of throwing heads with an unbiased coin is 1/2, because this is how unbiasedness is defined: each possible outcome is equally probable or equally likely. If, however, it is unknown whether the coin is unbiased, then it is not clear how the principle of indifference should be applied. A related objection, expressed by von Mises ([1928] 1981, pp. 69–70), is that it is often unclear what is meant by equally likely:

> according to certain insurance tables, the probability that a man forty years old will die within the next year is 0.011. Where are the 'equally likely cases' in this example? Which are the 'favourable' ones? Are there 1000 different possibilities, eleven of which are 'favourable' to the occurrence of death, or are there 3000 possibilities and thirty-three 'favourable' ones? It would be useless to search the textbooks for an answer, for no discussion on how to define equally likely cases in questions of this kind is given.

Von Mises concludes that, in practice, the notion of *a priori* 'equally likely' is substituted by some known long-run frequency. This step is unwarranted unless the foundation of probability also changes from indifference (or lack of knowledge) to frequencies (or abundance of knowledge) in a very precise setting in which a proper account of randomness is given (see the following chapter).

A second objection arises if the experiment is slightly complicated, by tossing the coin twice. What is the probability of throwing two heads? One solution, that works with partitions, would be 1/3, for there are three possibilities: two heads, two tails and a head and a tail. Today, this answer is considered wrong: the correct calculation should be based on the possible number of permutations. The mistake (or perhaps one should say 'different understanding') has been made by eminent scholars (for example Leibniz, see Hacking, 1975, p. 52), but even today it is not always clear how to formulate the set of equally possible alternatives (Hacking mentions a more recent but similar problem in small particle physics, where in some cases partitions, and in other cases permutations, have proved to be the relevant notion).

Third, the principle of indifference implies that, if there is no reason to attribute unequal probabilities to different events, the probabilities must be equal. This generates uniform probability distributions – in fact all probability statements based on the indifference theory are ultimately reduced to applications of uniform distributions. But again, what should be done if all that is known is that a coin is biased, without knowing in which direction?

Fourth, the application of a uniform distribution gives rise to paradoxes such as Bertrand's paradox (named after the mathematician Joseph Bertrand who discussed this paradox in 1889; see Keynes, [1921] CW VIII, p. 51; van Fraassen, 1989, p. 306). Consider figure 1, where an equilateral triangle ABC is inscribed in a circle with radius OR. What is the probability that a random chord XY is longer than one of the sides of the triangle? There are (at least) three alternative solutions to the problem if the principle of indifference is applied:

(i) Given that one point of the chord is fixed at position X on the circle. We are indifferent as to where on the circle the other end of the chord lies. If Y is located between 1/3 and 2/3 of the circumference away from X, then from the interval [0,1] the interval [1/3,2/3] provides the favourable outcomes. Hence the probability is 1/3.

(ii) We are indifferent as to which point Z the chord crosses orthogonally the radius of the circle. As long as OZ is less than half the radius, the chord is longer than the side of the triangle. Hence, the probability is 1/2.

(iii) We are indifferent with respect to the location of the middle point of the chord. In the favourable cases, this point lies in the concentric circle with half the radius of the original circle. The area of the inner circle is 1/4 of the area of the full circle, hence the probability is 1/4.

Obviously, the solutions are contradictory. This is one of the most persuasive arguments that has been raised against the indifference theory of

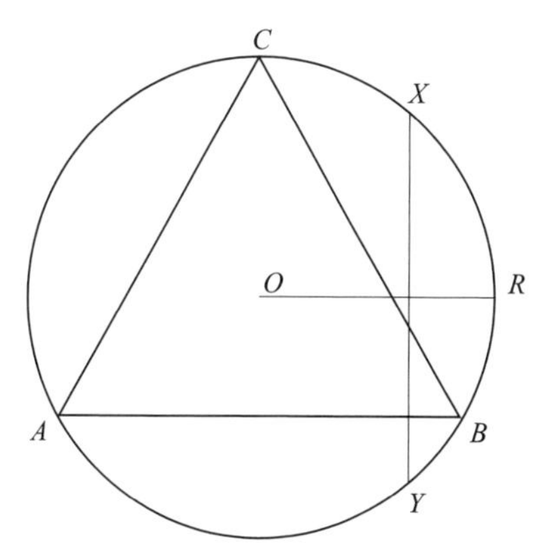

Figure 1

probability (see also the discusion of prior probability in chapter 4, section 5). Von Mises gives another version of this paradox, which deals with the problem of inferring the ratio of wine and water in a glass that contains a wine and water mixture. He concludes that the principle of indifference cannot be sustained. However, Jaynes (1973) shows that problems such as Bertrand's do have non-arbitrary solutions, if careful thought is given to all the invariance requirements involved. The greatest problem with the indifference argument is that it needs a lot of thought before it can be applied, and there may be cases where paradoxes will remain unsolved.

3 The rule of succession

How does the indifference theory of probability provide a basis for inductive reasoning? The probability *a priori*, based on insufficient reason, is mapped into a probability *a posteriori* by Laplace's *rule of succession*. No other formula in the alchemy of logic, Keynes ([1921] CW VIII, p. 89) writes, has exerted more astonishing powers. The simplest version of this rule is as follows (for a formal derivation, see Jeffreys, [1939] 1961, pp. 127–8; Cox and Hinkley, 1974, p. 368). If there is evidence from a sample of size n in which m elements have the characteristic C, we may infer from this evidence that the probability of the next element having characteristic C is equal to

$$P(C) = \frac{m+1}{n+2}. \qquad (2)$$

An interesting implication of the rule of succession is that sampling will only yield a high probability that a population is homogeneous if the sample constitutes a large fraction of the whole population (Jeffreys, [1939] 1961, p. 128). This is an unpleasant result for those with inductive aims (homogeneity of outcomes is an important condition for frequency theorists, like von Mises, and for proponents of epistemological probability, like Keynes and Jeffreys; see chapters 3 and 4). It results from using a uniform prior probability distribution to numerous (or infinite) possible outcomes. Keynes' principle of limited independent variety (see chapter 4, section 2) and the Jeffreys–Wrinch 'simplicity postulate' (discussed in chapter 5) are responses to this problem.

The validity of the rule of succession is controversial. Venn, Bertrand and Keynes reject the rule, while authors such as Jevons and Pearson support it.[1] Carnap generalizes the rule (see chapter 4, section 3.2). If n increases, the probability converges to the relative frequency, which yields the 'straight rule'. This rule sets the probability equal to the relative frequency.[2] On the other hand, if $n = 0$, then we may conclude that the probability that a randomly chosen man is named Laplace is equal to 1/2. The rule of succession raises the issue as to how to classify events in equiprobable categories.

Relying on the principle of indifference is likely to yield suspect inferences in economics. Take, for example, the dispute about the source of long-term unemployment: heterogeneity versus hysteresis. Heterogeneity means that a person who is long-term unemployed is apparently someone of low quality. Hysteresis means that unemployment itself reduces the chance of obtaining a new job, because the unemployed person loses human capital. Are these hypotheses *a priori* equiprobable? Perhaps, but it is very difficult to give a satisfactory justification for this choice (but see chapter 4). Moreover, how may evidence for different countries be used in a rule of succession? In Britain, hysteresis may provide the best explanation for unemployment spells, in the USA it may be heterogeneity. But the evidence is not beyond doubt, and inserting the cases in a rule of succession would be meaningless.

4 Summary

Only in rare cases, one can establish *a priori* whether events are equally likely. When this is not possible, the indifference theory of probability may result in paradoxes, notably Bertrand's. A related problem arises with respect to classification of events. Still, the indifference approach has been valuable as it inspired von Mises to his frequency theory of probability, and Keynes to his epistemological theory.

As an inductive principle, the method of indifference has yielded Laplace's rule of succession. This rule has clear drawbacks, but still figures in the background of some more advanced interpretations of probability, such as the one proposed by Carnap. After an intermezzo, I will now turn to these alternative interpretations: the frequency theory (chapter 3) and the epistemological theory (chapter 4).

Notes

1. The rule of succession is discussed in Keynes ([1921] CW VIII, chapter 30). Keynes' quibble about Laplace's calculations of the probability of a sunrise (see chapter 1 footnote 3, and Keynes, pp. 417–18) is the result of an abuse of this rule: the elements in the sample should be independent, which is clearly not the case here. Other discussions can be found in Fisher (1956, p. 24), Jeffreys ([1939] 1961, p. 127) and Howson and Urbach (1989, pp. 42–4).
2. Cohen (1989, p. 96) attributes the straight rule to Hume. Jeffreys ([1939] 1961, p. 370) attributes it to De Moivre's *Doctrine of Chances* (the first edition of this work appeared in 1718, when Hume was six years old). Jeffreys (p. 369) suggests that, more recently, Jerzy Neyman supports the straight rule, and criticizes Neyman. Neyman's (1952, pp. 10–12) reply shows that Jeffreys' suggestion is valid, but not all of his comments are.

I Intermezzo: a formal scheme of reference

In this intermezzo several basic concepts of probability are explained. Knowledge of these concepts is necessary for a good understanding of the remainder of the book. Readers who have such knowledge already may skip this intermezzo without a loss of understanding.

1 Probability functions and Kolmogorov's axioms

A. N. Kolmogorov provided the currently still-dominating formalism of probability theory based on set theory. Kolmogorov (1933, p. 3) interprets his formalism along the lines of von Mises' frequency theory, but the same formalism can be used for other interpretations (a useful introduction is given by Mood, Graybill and Boes, 1974, pp. 8–40, on which much of this section is based; a short discussion of the relation between sets and events can be found in Howson and Urbach, 1989, pp. 17–19).

First, the notions of sample space and event should be defined. The sample (or outcome) space S is the collection of all possible outcomes of a conceptual experiment.[1] An event A is a subset of the sample space. The class of all events associated with a particular experiment is defined to be the event space, $\mathcal{A}$. S is also called the sure event. The empty set is denoted by ϕ.

A probability function (or measure) $P(\cdot)$ is a set function with domain $\mathcal{A}$ and counterdomain the interval on the real line, [0, 1], which satisfies the following axioms (Kolmogorov, 1933, pp. 2 and 13):

(1) $\mathcal{A}$ is an algebra of events

(2) $\mathcal{A} \subset S$

(3) $P(A) \geq 0$ for every $A \in \mathcal{A}$

(4) $P(S) = 1$

(5) If $A_1, A_2, \ldots, A_n$ is a sequence of mutually exclusive events in $\mathcal{A}$

$$\left(\text{i.e. } \bigcap_{i=1}^{n} A_i = \phi\right)$$

and if $A_1 \cup A_2 \cup \ldots \cup A_n = \bigcup_{i=1}^{n} A_i \in \mathcal{A}$,

$$\text{then } P\left(\bigcup_{i=1}^{n} A_i\right) = \sum_{i=1}^{n} P(A_i).$$

(6) If $A_1, A_2, \ldots A_n$ is a sequence of mutually exclusive events in $\mathcal{A}$ and if $A_1 \supset A_2 \supset \ldots \supset A_n$,

$$\text{then } \lim_{n \to \infty} P(A_n) = 0.$$

The axioms have been challenged by various probability theorists (such as von Mises and De Finetti). Others thought them of little interest (for example, R. A. Fisher). For the purpose of this book, they are relevant, precisely because various authors have different opinions of their meaning. I will briefly discuss them below.

Defining the event space as an algebra (or *field*) implies that if A_1 and A_2 are events (i.e. they belong to $\mathcal{A}$), then also generalized events like $A_1 \cup A_2$ and $A_1 \cap A_2$ belong to $\mathcal{A}$. The use of algebras makes application of measure theory possible. Most of the interesting concepts and problems of probabilistic inference can be explained using probability functions without using measure theory.[2]

The second axiom states that the sample space contains the event space. A direct result of axiom 1 and 2 is that the empty set, ϕ, is an event as well. The third axiom restricts the counterdomain of a probability function to the non-negative real numbers. The fourth axiom states that the sure event has probability one. Some authors (e.g. Jeffreys, [1939] 1961) do not accept this axiom – they hold it as a convention which is often useful but sometimes is not. The fifth axiom is that of (finite) additivity, which is the most specific one for a probability function. Axiom 6, perhaps the most controversial of these axioms (rejected, for example, by de Finetti), is called the continuity axiom and is necessary for dealing with infinite sample spaces. These fifth and sixth axioms jointly imply countable additivity. It is easy to construct a sequence of events that generates a relative frequency, which is not a countable additive (see e.g. van Fraassen, 1980, pp. 184–6, for discussion and further references).

2 Basic concepts of probability theory

2.1 Probability space and conditional probability

The triplet $(\mathcal{S}, \mathcal{A}, P(\cdot))$ is the so-called *probability space*. If A and B are two events in the event space $\mathcal{A}$, then the *conditional probability* of A given B, $P(A|B)$, is defined by

$$P(A|B) = \frac{P(A) \cap B)}{P(B)} \qquad \text{if } P(B) > 0. \tag{I.1}$$

2.2 Bayes' theorem, prior and posterior probability

Combining (I.1) with a similar expression for $P(B|A)$ yields *Bayes' Theorem*:

$$P(A|B) = \frac{P(B|A)P(A)}{P(B)}. \tag{I.2}$$

In fact, the English Reverend Thomas Bayes was not the one who authored this theorem – his essay, posthumously published in 1763 (see Kyburg and Smokler, [1964] 1980), contains a special case of the theorem. A less restricted version (still for equiprobable events) of the theorem can be found in Pierre Simon Marquis de Laplace's memoir on 'the probability of causes', published in 1774. The general version of Bayes' theorem was given in the *Essay Philosophique* (Laplace [1814] 1951; see also Jaynes, 1978, p. 217). I will stick to the tradition of calling (I.2) Bayes' theorem. It is also known as the *principle of inverse probability*, as a complement to the *direct probability*, $P(B|A)$. Sometimes, a direct probability is regarded as the probability of a cause of an event. The indirect probability infers from events to causes (see Poincaré, [1903] 1952, chapter 11). The relation between probability and causality will be discussed in chapter 9, section 4.

A few remarks can be made. If an event is *a priori* extremely improbable, i.e. $P(A)$ approaches zero, then no finite amount of evidence can give it credibility. If, on the other hand, an event B is very unlikely given A, i.e. $P(B|A)$ approaches zero, but is observed, then A has very low probability *a posteriori*. These considerations make the theorem a natural candidate for analysing problems of inference. The theorem as such is not controversial. Controversial is, however, what is known as Bayes' Rule or Axiom to represent prior ignorance by a uniform probability distribution. The elicitation of prior probabilities is the central problem of Bayesian inference.

2.3 Independence and random variable

A few more concepts need to be defined. One is independence. Events A and B are *stochastically independent* if and only if $(P(A \cap B) = P(A)P(B)$. Hence, if A and B are independent, then $P(A|B) = P(A)$.

Until now, the fundamental notion of randomness has not been used. It seems odd that probability functions can be defined without defining this concept, but in practice, this difficult issue is neglected (an example is the otherwise excellent book of Mood, Graybill and Boes, 1974). I will come back to the notion of randomness later (see the discussions of von Mises, Fisher and de Finetti). For the moment, we have to assume that everybody knows, by intuition, what is meant by randomness. Then we can go on and define a random variable.

It is impossible to apply probability calculus to events like the occurrence of red, yellow and blue if we cannot assign numbers to these characteristics, and impose an ordering on the events $\mathcal{S} = \{\text{red, yellow, blue}\}$. The notion of a 'random variable' serves this purpose. If the probability space $(\mathcal{S}, \mathcal{A}, P(\cdot))$ is given, then the *random variable* $X(\cdot)$ is a function from $\mathcal{S}$ to $\mathbb{R}$. Every outcome of an experiment now corresponds with a number. Mood, Graybill and Boes (1974) acknowledge that there is no justification for using the words 'random' and 'variable' in this definition.

2.4 Distribution and density functions

The *cumulative distribution function* of X, $F_X(\cdot)$ is defined as that function with domain $\mathbb{R}$ and counterdomain $[0, 1]$ which satisfies $F_X(x) = P(\omega{:}X(\omega \leq x))$ for every real number x. If X is a continuous random variable, i.e. if there exists a function $f_X(\cdot)$ such that

$$F_X(x) = \int_{-\infty}^{x} f_X(u)du \qquad \text{for every } x, \tag{I.3}$$

then the integrand $f_x(\cdot)$ in $F_X(x) = \int_{-\infty}^{x} f_X(u)du$ is called the *probability density function* for this continuous random variable. A geometric interpretation is that the area under $f_X(\cdot)$ over the interval (a, b) gives the probability $P(a < X < b)$. Similar notions exist for discrete random variables. It is also possible to extend the definition to jointly continuous random variables and density functions. The k-dimensional random variable $(X_1, X_2, \ldots, X_k)$ is a continuous random variable if and only if a function $f_{X_1,\ldots X_k}(\cdot, \ldots, \cdot) \geq 0$ does exist such that

$$F_{X_1,\ldots,X_k}(x_1,\ldots,x_k) = \int_{-\infty}^{x_k} \cdots \int_{-\infty}^{x_1} f_{X_1,\ldots,X_k}(u_1,\ldots,u_k)du_1,\ldots,du_k, \tag{I.4}$$

for all $(x_1, \ldots, x_k)$. A joint probability density function is defined as any non-negative integrand satisfying this equation.

Finally, consider the concepts of marginal and conditional density functions. They are of particular importance for the theory of econometric modelling, outlined in chapter 7. If X and Y are jointly continuous random variables, then their *marginal probability density functions* are defined as

$$f_X(x) = \int_{-\infty}^{\infty} f_{X,Y}(x, y)dy \tag{I.5}$$

and

$$f_Y(y) = \int_{-\infty}^{\infty} f_{X,Y}(x, y)dx \tag{I.6}$$

respectively. The *conditional probability density function* of Y given $X = x$, $f_{Y|X}(\cdot|x)$ is defined

$$f_{Y|X}(\cdot|x) = \frac{f_{X,Y}(x, y)}{f_X(x)} \qquad \text{for } f_X(x) > 0. \tag{I.7}$$

These are the fundamental notions of probability which are frequently used in what follows. Note that some of the most fundamental notions, such as randomness and, in general, the interpretation of probability, have not yet been defined. Quine (1978) notes that mathematics in general (hence probability theory in particular) cannot do without a non-formal interpretation. Chapters 2, 3 and 4 deal with the interpretation of different theories of probability.

Notes

1. The concept of a sample space seems to be used first by R. A. Fisher in a 1915 *Biometrika* article on the distribution of the correlation coefficient. The notion of a conceptual experiment is due to Kolmogorov, and is defined as any partition of S, obeying some elementary set-theoretic requirements (see Kolmogorov, 1933).
2. See Chung (1974) or Florens, Mouchart and Rolin (1990) for a more elaborate discussion of the technical issues and refinements of the axioms.

3 Relative frequency and induction

> They wanted facts. Facts! They demanded facts from him, as if facts could explain anything!
>
> Joseph Conrad, *Lord Jim*

1 Introduction

The frequentist interpretation of probability dominates mainstream probability theory and econometric theory. Proponents of the frequency theory bolster its alleged objectivity, its exclusive reliance on facts. For this reason, it is sometimes called a 'realistic' theory (Cohen, 1989).

Keynes ([1921] CW VIII, p. 100) pays tribute to the English mathematician Leslie Ellis, for providing the first logical investigation of the frequency theory of probability. Ellis (1843) objects to relating probability to ignorance: *ex nihilo nihil*. John Venn ([1866] 1876) elaborated the frequency theory. Still, a theory which is formally satisfactory did not exist until the early twentieth century.[1] In 1919, Richard von Mises formulated the foundations for a frequency theory of probabilistic inference (see von Mises, [1928] 1981, p. 224). His views are discussed in, section 2. Shortly after, Ronald Aylmer Fisher (1922) presented an alternative frequentist foundation of probabilistic inference. Section 3 deals with Fisher's theory. Today, the most popular frequentist theory of probability is based on yet another interpretation, due to Jerzy Neyman and Egon Pearson. This is the topic of, section 4. Popper's interpretation of probability, which is supposed to be entirely 'objective', is discussed in section 5. Section 6 offers a summary of the chapter.

I will argue that, despite the elegance of the theory of von Mises and the mathematical rigour of the Neyman–Pearson theory, those theories have fundamental weaknesses which impede econometric inference. The framework proposed by Fisher comes closer to the needs of econometrics, while Popper's theory is of no use.

2 The frequency theory of Richard von Mises

2.1 The primacy of the collective

The principal goal of von Mises was to make probability theory a science similar to other sciences, with empirical knowledge as its basis. He is a follower of the positivist scientist and philosopher Ernst Mach. In line with Ellis, von Mises ([1928] 1981, p. 30) criticizes the view that probability can be derived from ignorance:

> It has been asserted – and this is no overstatement – that whereas other sciences draw their conclusions from what we know, the science of probability derives its most important results from what we do not know.

Rather, probability should be based on facts, not their absence. Indeed, the frequency theory relates a probability directly to the 'real world' via the observed 'objective' facts (i.e. the data), preferably repetitive events. If people position the centre of gravity of an unknown cube in its geometric centre, they do not do so because of their lack of knowledge. Instead, it is the result of actual knowledge of a large number of similar cubes (p. 76). Probability theory, von Mises (1964, p. 98) writes, is an empirical science like other ones, such as physics:

> Probability theory as presented in this book is a mathematical theory and a science; the probabilities play the role of physical constants; to any probability statement an approximate verification should be at least conceivable.

Von Mises' (pp. 1 and 13–14) probability theory is not suitable for investigating problems such as the probability that the two poems *Illiad* and *Odyssey* have the same author. There is no reference to a prolonged sequence of cases, hence it hardly makes sense to assign a numerical value to such a conjecture. The authenticity of Cromwell's skull, investigated by Karl Pearson with statistical techniques, cannot be analysed in von Mises' probabilistic framework. More generally, there is no way to 'probabilify' scientific theories.

This distinguishes his frequency theory from the one of his close colleague, Reichenbach. According to von Mises, a theory is right or false. If it is false, then it is not a valid base for probabilistic inference. Probability should not be interpreted in an epistemological sense. It is not lack of knowledge (uncertainty) which provides the foundation of probability theory, but experience with large numbers of events. Like von Mises, Reichenbach ([1938] 1976, p. 350) holds that the aim of induction '*is to find series of events whose frequency of occurrence converges toward a limit*'. Both argue that probability is a frequency limit. However, Reichenbach (1935, p. 329) claims that it is perfectly possible to determine the probability of the quantum theory of physics being correct. It is even possible to fix a numerical probability. This can be done, if

man sie mit anderen physikalischen Theorien ähnlicher Art in eine Klasse zusammenfaßt und für diese die Wahrscheinlichkeit durch Auszählung bestimmt

('if it is put in a class with other similar physical theories and one determines their probability by counting' i.e. one should construct an urn with similar theories of physics). These probabilities are quantifiable, Reichenbach argues. How his urn can be constructed in practice, and how the probability assignments can be made, remains obscure (except for the suggestion that one may formulate a bet which may yield a consistent foundation of probabilistic inference, but this implies a departure from the basic philosophy underlying a frequentist theory of probability). Therefore, Reichenbach's theory is of little help.

The basis of von Mises' theory is the *collective*. A collective is a sequence of uniform events or processes, which differ in certain observable attributes such as colours, numbers or velocities of molecules in a gas (von Mises, [1928] 1981, p. 12). It must be defined before one can speak of a probability: first the collective, then the probability. A collective has to fulfil two conditions, the convergence condition and the randomness condition. They are defined as follows.

Convergence condition. Let the number of occurrences of an event with characteristic i be denoted by n_i, and the number of all events by n. Then

$$\lim_{n \to \infty} \frac{n_i}{n} = p_i, \qquad i = 1, 2, \ldots, k, \tag{1}$$

where p_i is the limiting frequency of characteristic i.

Randomness condition. Let m be an infinite sub-sequence of sequence n, derived by a place selection. A place selection is a function ϕ that selects an element out of sequence n where the selection criterion may depend on the value of already selected events, but does not depend on the value of the event to be selected or subsequent ones in n. Then it should be the case that

$$\lim_{m \to \infty} \frac{m_i}{m} = p_i, \qquad i = 1, 2, \ldots, k. \tag{2}$$

The convergence condition implies that the relative frequencies of the attributes must possess limiting values (this is an elaboration of the Venn limit). The randomness condition implies that the limiting values must remain the same in all arbitrary sub-sequences (von Mises, [1928] 1981, pp. 24–5). The randomness condition is also known as the principle of the impossibility of a gambling system (p. 25). Fisher (1956, p. 32) proposes a condition of 'no recognizable subsets' which is similar to the randomness condition. An example in time-series econometrics is the criterion of serially uncorrelated residuals of a regression.

Von Mises (1964, p. 6) notes that an observed frequency may be different from its limiting value, for example, a specific sequence of 1,000,000 throws of a die may result in observing only sixes even if its limiting frequency is 1/6. Indeed,

> [i]t is a silent assumption, *drawn from experience that in the domain where probability calculus has (so far) been successfully applied*, the sequences under consideration do not behave this way; they rather exhibit *rapid convergence.*

Note that the convergence condition is not a restatement of the law of large numbers (a mathematical, deductive theorem – see below). It is postulated on the basis of empirical observation. The same applies to the randomness condition (von Mises, [1928] 1981, p. 25).

Formally von Mises should have made use of a probability limit rather than a mathematical limit in his definition of the convergence condition. In that case it becomes problematic to define the collective first, and then probability. Like the indifference interpretation, the frequency interpretation uses a circular argument. However, von Mises is an empiricist who bases his interpretation of probability on empirical statements, not on formal requirements. Moreover, if von Mises were able to define a random event without making use of a primitive notion of probability, his program (first the collective, then probability) might be rescued.

2.2 *Inference and the collective*

How might von Mises' probability theory be used for scientific inference or induction? The answer to this question cannot be given before the law of large numbers is discussed. It dates back to Bernoulli, has been coined and reformulated by Poisson and is one of the cornerstones of probability theory since. Without success, Keynes ([1921] CW VIII, p. 368) suggested renaming this law the 'stability of statistical frequencies', which provides a clear summary of its meaning.

There are different versions of this law. Chebyshev's 'weak law of large numbers' follows.[2]

'Weak law of large numbers' (Chebyshev). Let $f(\cdot)$ be a probability density function with mean μ and (finite) variance σ^2, and let x_n be the sample mean of a random sample of size n from $f(\cdot)$. Let ϵ and δ be any two specified numbers satisfying $\epsilon > 0$ and $0 < \delta < 1$. If n is any integer greater than $\sigma^2/\epsilon^2\delta$, then

$$P(|x_n - \mu| < \epsilon) \geq 1 - \delta. \qquad (3)$$

This law has a straightforward interpretation. If the number of observations increases, the difference between the sample mean and its 'true' value μ becomes arbitrarily small. In other words (von Mises, 1964, p. 231), for large n it is almost certain that the sample mean will be found in the immediate vicinity of its expected value. This seems to provide a 'bridge' (a notion used by von Mises, [1928] 1981, p. 117; also used by Popper, [1935] 1968) between observations and probability. But the law of large numbers is a one-way bridge, from the 'true' value of the probability to the observations.[3]

Is inference in the opposite direction possible, from the sequence of observations to the expected value? The oldest answer to this question is the straight rule (mentioned in chapter 2) which states that the probability is equal to the observed frequency. This rule presupposes a very specific prior probability distribution which is not generally acceptable. The recognition of this point seems to have been the basis for Laplace's rule of succession. Von Mises ([1928] 1981, p. 156; 1964, p. 340) proposes another solution: to combine the law of large numbers with Bayes' theorem (see the Intermezzo). He calls the result the *second law of large numbers* (the first being Bernoulli's, a special case of Chebyshev's; see Spanos, 1986, pp. 165–6).[4] This second law is formulated such that the numerical value of the prior probabilities or its specific shape is irrelevant, except that the prior density $p_0(x)$ (using the notation of von Mises, 1964, p. 340) is continuous and does not vanish at the location of the relative frequency r, and furthermore, $p_0(x)$ has an upper bound. Under those conditions, one obtains the

'Second law of large numbers' (von Mises, 1964, p. 340). If the observation of an n-times repeated alternative shows a relative frequency r of 'success', then, if n is sufficiently large, the chance that the probability of success lies between $r - \epsilon$ and $r + \epsilon$ is arbitrarily close to one, no matter how small ϵ is.

The frequentist solution to the problem of inference given by von Mises consists of a combination of the frequency concept of a collective with Bayes' theorem. If knowledge of the prior distribution does exist, there is no conceptual problem with the application of Bayes' theorem. If, as is more often the case, this prior information is not available, inference depends on the availability of a large number of observations (von Mises, 1964, p. 342; emphasis added):

> It is not right to state that, in addition to a large number of observations, knowledge of [the prior distribution] is needed. *One or the other is sufficient.*

But von Mises underestimates some of the problems that arise from the application of non-informative prior probabilities or neglecting priors. In many cases, von Mises is right that the prior vanishes if the number of observations increases, but this is not always true. In the case of time series inference, where non-stationarity is a possibility, the choice of the prior is crucial for the conclusion to be drawn, even with large numbers of observations (see for example Sims and Uhlig, 1991).

2.3 Appraisal

Von Mises' theory has been very useful to clarify some problems in probability theory. For example, it avoids Bertrand's paradox because it restricts probability to cases where we can obtain factual knowledge, for example a collective of water–wine mixtures. In that case, the limiting ratio of water and wine does exist, its value depends on the construction of the collective. Different constructions are possible, yielding different limiting frequencies. The problem is to consider, in an empirical context, first, whether a sequence satisfies the conditions of a collective, and if so, to infer its probability distribution.

Obviously, a practical difficulty arises as a collective is defined for an *infinite* sequence. A collective is an idealization. Von Mises ([1928] 1981, p. 84) acknowledges that '[i]t might thus appear that our theory could never be tested empirically'. He compares this idealization with other popular ones, such as the determination of a specific weight (perfect measurement being impossible) or the existence of a point in Euclidean space. The empirical validity of the theory does not depend on a logical solution, but is determined by a practical decision (von Mises, p. 85). This decision should be based on previous experience of successful applications of probability theory, where practical studies have shown that frequency limits are approached comparatively rapidly. Von Mises (1964, p. 110) again stresses the empirical foundation of probability:

> *Experience has taught us* that in certain fields of research these hypotheses are justified, – we do not know the precise domain of their validity. In such domains, and only in them, statistics can be used as a tool of research. However, in 'new' domains where we do not know whether 'rapid convergence' prevails, 'significant' results in the usual sense may not indicate any 'reality'.

This may be a weak foundation for an objectivist account of probability, but von Mises deserves credit for its explicit recognition. If the data are not regular in the sense of his two conditions, then statistical inference lacks a foundation and Hume's problem remains unresolved. Von Mises' probabilistic solution to Hume's problem is, after all, a pragmatic one.

Apart from the practical problem of identifying a finite set of observations with the infinite notion of a collective, von Mises' theory suffers from some formal weaknesses. Unlike the formalist Kolmogorov, von Mises provides a very precise *interpretation* of the notion of probability and randomness. Kolmogorov takes probability as a primitive notion, whereas von Mises attempts to derive the meaning of probability from other notions, in particular convergence and randomness. Kolmogorov notes that the two approaches serve different ends: formal versus applied. Von Mises is interested in bridging the mathematical theory with the 'empirische Entstehung des Wahrscheinlichkeitsbegriffes' (i.e. the empirical origin of the notion of probability; Kolmogorov, 1993, p. 2). For this purpose, Kolmogorov (p. 3, n. 1) accepts the empirical interpretation of von Mises.

The formal circularity of the argument, mentioned in section 2.1, has rarely been criticized. Instead, von Mises' interpretation has been attacked by followers of Kolmogorov for an alleged conceptual weakness, related to the notion of place selection. Von Mises emphasizes the constructive role of a place selection in his randomness condition – there are no restrictions to the resulting sub-sequence itself. But formally, it can be proved that there exists a place selection which provides an infinite sequence of random numbers, in which only one number repeatedly occurs.[5] Is this still a 'random sequence'? Von Mises does not care. To understand why, it is useful to distinguish between different approaches in mathematical inference. Most mathematicians today prefer to build mathematical concepts from *formal structures* (sets of relations and operations) and axioms to characterize the structures ('Nicolas Bourbaki', a pseudonym for a group of French mathematicians, advanced this approach in a series of volumes starting from 1939; Kolmogorov was a forerunner). Another school (headed by L. E. J. Brouwer) only accepts *intuitive concepts*. *Intuitionism* holds that mathematical arguments are mental constructs. Von Mises, finally, follows a *constructive* approach (intuitionism and finitism are special versions thereof). He argues that constructivism makes his theory empirical (scientific, like physics). The formal possibility that a very unsatisfactory sequence may be constructed is unlikely to be realized in practice, given an acceptable place selection (Von Mises, [1928] 1981, pp. 92–93). Abraham Wald (1940) showed that if the set of place selections $\{\phi\}$

is restricted to a countable set, the problem disappears. In the same year Church proposed identifying this set with the set of recursive (i.e. effectively computable) functions. The Wald–Church solution is a step forward but not yet perfect (see Li and Vitányi, 1990, p. 194). A final refinement has been proposed by replacing 'first the collective, then probability' by 'first a random sequence, then probability'. This approach will be pursued in the chapter 4, section 4.

Von Mises is rarely cited in econometrics (Spanos, 1986, p. 35, is an exception; another exception is the review of von Mises, [1928] 1981, by the philosopher Ernest Nagel in *Econometrica,* 1952). His influence is not negligible, though. As shown above, Kolmogorov supports von Mises' interpretation of probability. Similarly, the probability theorist Jerzy Neyman (1977, p. 101), who bases his methods primarily on Kolmogorov formalism, is 'appreciative of von Mises' efforts to separate a frequentist probability theory from the intuitive feelings of what is likely or unlikely to happen'. Many of von Mises' ideas entered econometrics indirectly, via his student Wald. Harald Cramér (1955, p. 21) constructed a probability theory which he claims is closely related to the theory of von Mises. This small group of people inspired Trygve Haavelmo and other founders of formal econometrics.

If the theory of von Mises is at all applicable to economics, it must be in cross-section econometrics where data are relatively abundant. However, it is an open question whether the criterion of convergence is satisfied in micro-econometric applications. Probabilities in economics are not the kind of physical entities that von Mises seems to have in mind in constructing his theory. Although I do not have direct evidence for this proposition, this may be a reason why von Mises' brother, the economist Ludwig von Mises, rejects the use of formal probability theory in economics. Even so, the subtlety of his work justifies a serious attempt at an interpretation of econometric inference as the study of social (rather than physical) collectives.

Von Mises' effort to provide a positivist (empirical) foundation for the frequency theory of probability is laudable. But the strength of the theory is also a weakness: it is impossible to analyse probabilities of individual events with a theory based on collectives, or to perform small sample analysis without exact knowledge of prior probabilities. Here, the work of R. A. Fisher comes in.

3 R. A. Fisher's frequency theory

Sir Ronald Aylmer Fisher was a contemporary of von Mises. Both developed a frequency theory of probability. Fisher was influenced by Venn, the critic of 'inverse probability'or Bayesian inference, and President of Gonville and Caius College whilst Fisher was an undergraduate student there (Edwards, [1972] 1992, p. 248). While von Mises was a positivist in the tradition of Mach, Fisher was an exponent of British empiricism, often associated with the names of Locke, Berkeley and Hume. Fisher's methodological stance is more instrumentalistic than von Mises'. The latter excels in analytic clarity. Fisher, on the other hand, is characterized by intuitive genius, less by analytical rigour. His theoretical conjectures have often been proved by other statisticians, but he also succeeded in making empirical conjectures (predicting 'new facts' which were verified afterward).[6] Fisher is the man of the Galton laboratory and Rothamstead Experimental Station, where theory met practical application, which he considered to be more important for the development of the theory of probability than anything else. His innovations in the vocabulary and substance of statistics have made him immortal.

3.1 Fisher on the aim and domain of statistics

In his classic paper on the foundations of statistics, Fisher (1922, p. 311) gives the following description of the aim of statistics:

> the object of statistical methods is the reduction of data. A quantity of data, which usually by its mere bulk is incapable of entering the mind, is to be replaced by relatively few quantities which shall adequately represent the whole, or which, in other words, shall contain as much as possible, ideally the whole, of the relevant information contained in the original data.
>
> This object is accomplished by constructing a hypothetical infinite population, of which the actual data are regarded as constituting a random sample.

Reduction serves induction, which is defined as 'reasoning from the sample to the population from which the sample was drawn, from consequences to causes, or in more logical terms, from the particular to the general' (Fisher, 1955, p. 69). Consistency, efficiency, sufficient and ancillary statistics, maximum likelihood, fiducial inference are just a few of Fisher's inventions of new terms for data reduction. Fisher is well aware, however, that a theory of data reduction by means of estimation and testing is not yet the same as a theory of scientific inference. He accepts Bayesian inference if precise prior information is available, but notes that in most cases prior probabilities are unknown. For these circumstances, he proposes the fiducial argument, 'a bold attempt to make the Bayesian omelet without breaking the Bayesian eggs' (Savage, 1961, p. 578). I will first discuss a number of elements of Fisher's theory of probability, and then ask whether econometricians may use the fiducial argument for inference.

Unlike von Mises, Fisher does not restrict the domain of probability theory to 'collectives'. Fisher takes the hypothetical infinite population (introduced in Fisher, 1915) as the starting point of the theory of probability. This resembles the *ensemble* of Willard Gibbs, who used this notion in his theory of statistical mechanics. A hypothetical infinite population seems similar to von Mises' collective, but there are two differences. First, von Mises starts with observations, and then postulates what happens if the number of observations goes off to infinity. Fisher starts from an infinite number of possible trials and defines the probability of A as the ratio of observations where A is true to the total number of observations. Secondly, unlike von Mises, Fisher has always been interested in small sample theory. Those who are not, Fisher (1955, p. 69) claims, are 'mathematicians without personal contact with the Natural Sciences'.[7] Fisher was proud of his own laboratory experience, and regarded this as a hallmark of science.

According to Fisher (1956, p. 33), it is possible to evaluate the probability of an outcome of a single throw of a die, if it can be seen as a random sample from an aggregate, which is 'subjectively homogeneous and without recognizable stratification'. Before a limiting ratio can be applied to a particular throw, it should be subjectively verified that there are no recognizable subsets. This condition is similar to von Mises' condition of randomness, although the latter is better formalized and, of course, strictly intended for investigating a collective of throws, not a single throw.

Another difference between the theories of von Mises and Fisher is that, in Fisher's theory, there is no uncertainty with respect to the parameters of a model: the 'true' parameters are fixed quantities, the only problem is that we cannot observe them directly. Von Mises ([1928] 1981, p. 158) complains that he does

> not understand the many beautiful words used by Fisher and his followers in support of the likelihood theory. The main argument, namely, that p is not a variable but an 'unknown constant', does not mean anything to me.

Moreover, von Mises (1964, p. 496) argues:

> The statement, sometimes advanced as an objection [to assigning a prior probability distribution to parameter θ], that θ is not a variable but an 'unknown constant' having a unique value for the coin under consideration, is beside the point. Any statement concerning the chance of θ falling in an interval H (or concerning the chance of committing an error, etc.) is necessarily a statement about a universe of coins, measuring rods, etc., with various θ-values.

For the purpose of inference, the frequentist asserts implicitly, or explicitly, that the statistical model is 'true' (Leamer, 1978, calls this the 'axiom of correct specification'). This is a strong assumption if two competing theories, which both have *a priori* support, are evaluated![8] Fisher did not discuss the problem of appraising rival theories, but he was fully aware of the specification problem. Indeed, one of his most important contributions to statistics deals with the problem of experimental design, which has a direct relation to the specification problem (which features prominently in chapters 6 and 7). First, a number of Fisher's innovations to the statistical box of tools should be explained, followed by other theories of inference. Fisher's tools have obtained wide acceptance, whereas his theory of inference has not.

3.2 Criteria for statistics

Like Hicks did for economics, Fisher added a number of most useful concepts to the theory of probability, i.e. consistency, efficiency, (maximum) likelihood, sufficiency, ancillarity, information and significance test.

Consistency

According to Fisher (1956, p. 141), the fundamental criterion of estimation is the Criterion of Consistency. His definition of a consistent statistic is: 'A function of the observed frequencies which takes the exact parametric value when for these frequencies their expectations are substituted' (p. 144).[9] This definition of consistency is actually what today is known as unbiasedness (I will use the familiar definitions of consistency and unbiasedness from now on). It would reject the use of sample variance as an estimator of the variance of a normal distribution. A condition, which according to Fisher (p. 144) is 'much less satisfactory', is *weak* consistency (today known as plain consistency): the probability that the error of estimation of an estimator t_n for a parameter θ given sample size n converges to zero if n grows to infinity ($\text{plim}_{n \to \infty} t_n = \theta$; a consequence of Bernoulli's weak law of large numbers and, in the theory of von Mises, a prerequisite for probabilistic inference – his condition of convergence). Fisher desires to apply his criterion to small samples (which is not possible if the asymptotic definition is used), therefore, he prefers his own definition.

The importance of unbiasedness and consistency is not generally accepted. Fisher regards unbiasedness as fundamental. Many authors accept consistency as obvious. Spanos (1986, p. 244) thinks consistency is 'very important'. Cox and Hinkley (1974, p. 287) give it 'a limited, but nevertheless important, role'. Efron (1986, p. 4) holds that unbiasedness is popular for its intuitive fairness. The other side of the spectrum is the Bayesian view, presented, for example, by Howson and Urbach (1989, p. 186) who see no merit in consistency as a desideratum, or by Savage ([1954] 1972, p. 244) who asserts that a serious reason to prefer unbiased estimates has never been proposed. The student of Fisher, C. R. Rao ([1965] 1973, p. 344), makes fun of consistency by defining the consistent estimator σ_n as $\sigma_n = 0$ if $n \leq 10^{10}$, or $\sigma_n = \theta_n$ if $n > 10^{10}$, where θ_n is a more satisfactory ('consistent' in its intended meaning) estimator of θ. In case of sample sizes usually available in economics, this suggest using 0 as the (consistent) estimate for any parameter θ. Minimizing a mean squared prediction error (given a specific time horizon) may be much more useful in practice than aiming at consistency.

One reason for the appeal of consistent or unbiased estimators may be that few investigators would be willing to assert that, in general, they prefer biased estimators because of the emotional connotations of the word 'bias'. If this is the rhetoric of statistics in practice, then this also applies to the notion of efficiency.

Efficiency

A statistic of minimal limiting variance is 'efficient' (Fisher, 1956, p.156). In the case of biased estimators, efficiency is usually defined as minimizing the expected mean square error, $\mathcal{E}(t_n - \theta)^2$. Efficiency 'expresses the proportion of the total available relevant information of which that statistic makes use' (Fisher, 1922, p. 310).

Maximum likelihood

The likelihood function, perhaps Fisher's most important contribution to the theory of probability, gives the likelihood that the random variables $X_1, X_2, \ldots, X_n$ assume particular values $x_1, x_2, \ldots, x_n$, where the likelihood is the value of the density function at a particular value of θ. The method of maximum likelihood is to choose the estimate for which the likelihood function has its maximum value. Fisher (1956, p. 157) argues that

> no exception has been found to the rule that among Consistent Estimates, when properly defined, that which conserves the greatest amount of information is the estimate of Maximum Likelihood.

He continues that the estimate that loses the least of the information supplied by the data must be preferred, hence maximum likelihood is suggested by the sufficiency criterion.

Maximum likelihood has a straightforward Bayesian interpretation (choosing the mode of the posterior distribution if the prior is non-informative). Without such a Bayesian or other bridge inductive inference cannot be based on just maximum likelihood. Wilson (cited in Jeffreys [1939] 1961, p. 383), states: 'If maximum likelihood is the only criterion the inference from the throw of a head would be that the coin is two-headed.' But, according to Jeffreys, in practical application to estimation problems, the principle of maximum likelihood is nearly indistinguishable from inverse probability. Recent investigations in time series statistics suggest that this is not true in general.

Sufficiency

The criterion of sufficiency states that a statistic should summarize the whole of the relevant information supplied by a sample. This means that, in the process of constructing a statistic, there should be no loss of information. This is perhaps the most fundamental requisite in Fisher's theory. Let $x_1, \ldots x_n$ be a random sample from the density $f(\cdot; \theta)$. Define a *statistic* T as a function of the random sample, $T = t(x_1, \ldots x_n)$, then this is a sufficient statistic if and only if the conditional distribution of $x_1, \ldots x_n$ given $T = t$ does not depend on θ for any value t of T (see Mood, Graybill and Boes 1974, p. 301; Cox and Hinkley, 1974, p. 18). This means that it does not make a difference for inference about θ if a sufficient statistic $T = t(x_1, \ldots x_n)$ is used, or $x_1, \ldots x_n$ itself. Cox and Hinkley (1974, p. 37) base their 'sufficiency principle' for statistical inference on this Fisherian notion. The econometric methodology of David Hendry is in the spirit of this principle, as will be shown in chapter 7, but other approaches to econometric inference hold sufficiency in high esteem as well. Sufficient statistics do not always exist (they do exist for the exponential family of distributions, such as the normal distribution). The Rao–Blackwell lemma states that sufficient statistics are efficient. A statistic is *minimal sufficient*, if the data cannot be reduced beyond a sufficient statistic T without losing sufficiency. A further reduction of the data will result in loss of information.

In 1962, Birnbaum showed that the principle of sufficiency, together with the principle of conditionality (see Berger, [1980] 1985, p. 31) yields the 'Likelihood Principle' as a corollary. This principle states that, given a statistical model with parameters θ and given data x, the likelihood function $l(\theta) = f(x|\theta)$ provides all the evidence concerning θ contained by the data. Furthermore, two likelihood functions contain the same information about θ if they are proportional to each other (see pp. 27–33, for further discussion). Fisher's theory of testing, discussed below, violates this likelihood principle.

Ancillary statistic

If S is a minimal sufficient statistic for θ and $\dim(S) > \dim(\theta)$ then it may be the case that $S = (T, C)$ where C has a marginal distribution which does not depend on θ. In that case, C is called an ancillary statistic, and T is sometimes called conditionally sufficient (Cox and Hinkley, 1974, pp. 32–3). Ancillarity is closely related to exogeneity (discussed in chapter 7).

Information

The second derivative of the logarithm of the likelihood function with respect to its parameter(s) θ provides an indication of the amount of information (information matrix) realized at any estimated value for θ (Fisher, 1956, p. 149). It is known as the Fisher information. Normally, it is evaluated at the maximum of the likelihood function. Fisher interprets the information as a weight of the strength of the inference. The Cramér–Rao inequality states that (subject to certain regularity conditions) the variance of an unbiased estimator is at least as large as the inverse of the Fisher information. Thus, the lower the variance of an estimator, the more 'information' it provides (it has, therefore, relevance for the construction of non-informative prior probabilities).

Significance test and *P*-value

An important notion for testing hypotheses is the *P*-value, invented by Karl Pearson for application with the χ^2 goodness-of-fit test. If χ^2 is greater than some benchmark level, then the tail-area probability P is called significantly small. The conclusion is that either the assumed distribution is wrong or a rare event under the distribution must have happened. 'Student' (Gosset) took the same approach for the t-test. Fisher continues this tradition. Fisher ([1935] 1966, p. 16) claims that the *P*-value is instrumental in inductive inference. Anticipating Popper, he argues that 'the null hypothesis is never proved or established, but is possibly disproved in the course of experimentation'. An experimental researcher should try to reject the null hypothesis. Unlike Popper, Fisher claims that this is the source of *inductive* inference. The value of an experiment increases whenever it becomes easier to disprove the null (increasing the size of an experiment and replication are two methods to increase this value of an experiment: replication is regarded as another way of increasing the sample size, p. 22). It is important to note that Fisher's approach not only differs from Popper's, in that Popper rejects all kinds of inductive arguments, but also that Fisher's null hypothesis is usually not the hypothesis the researcher likes to entertain.

Basing inference on P-values has a strange implication: the judgement is driven by observations that in fact are not observed. According to Jeffreys ([1939] 1961, p. 385) the P-value

> gives the probability of departures, measured in a particular way, equal to *or greater than* the observed set, and the contribution from the actual value is nearly always negligible. *What the use of P implies, therefore, is that a hypothesis that may be true may be rejected because it has not predicted observable results that have not occurred.* This seems to be a remarkable procedure.

Jeffreys anticipates the conclusion that inference based on P-values violates the 'likelihood principle'.

Fisher ([1925] 1973) suggests using significance levels of 1% or 5% and offers tables for a number of tests. The reason for this (and the subsequent popularity of those percentages) is a historical coincidence: due to his bad relations with Karl Pearson, Fisher was prevented from publishing Pearson's tables which contained probability values in the cells. Fisher had, therefore, to create his own way of tabulation which subsequently proved more useful (see Keuzenkamp and Magnus, 1995). The tables facilitate the communication of test results, but reporting a P-value provides more information than reporting that a test result is significant at the 5% level. Furthermore, Jeffreys' complaint about P-values applies equally well to Fisher's significance levels. Finally, Fisher does not give formal reasons for selecting one rather than another test method for application to a particular problem which is a considerable weakness as different test procedures may lead to conflicting outcomes. Neyman and Pearson suggested a solution for this shortcoming (see below).

3.3 Randomization and experimental design

As a practical scientist, Fisher paid much attention to real problems of inference. The titles of his books, *Statistical Methods for Research Workers* and *The Design of Experiments*, reflect this practical interest. The latter contains an elaborate discussion of one of Fisher's major contributions to applied statistics, randomization. Randomization provides 'the physical basis of the validity of the test' (the title of chapter 9 of Fisher [1935] 1966, p. 17). The 'inductive basis' for inferences, Fisher (p. 102) argues, is improved by varying the conditions of the experiment. This is a way to control for potentially confounding influences (nuisance variables) which may bias the inference with regard to the variables of interest (another option is to control directly, by means of including all potentially relevant and well measured variables in the multiple regression model; see also chapter 6, section 3 for discussion). Fisher also regards randomization as a means to justify distributional assumptions, thereby avoiding mis-specification problems (this justification works often, though not always).[10] Randomization, when introduced, was revolutionary and controversial (K. Pearson, for example, had developed his system of distributions in order to deal with all kinds of departures from normality, which after randomization became relatively unimportant; Fisher Box (1978) provides interesting historical background information).

A good introduction to experimental design can be found in Fisher ([1935] 1966). Here, the famous example of the 'tea-cup lady' (likely to be inspired by a female assistant at Rothamsted, where regular afternoon tea sessions took place), who claims to be able to detect whether tea or milk was put in her cup first (p. 11). In order to test the hypothesis, many variables may affect the outcome of the experiment apart from the order of milk and tea (e.g. thickness of the cup, temperature of the tea, smoothness of material etc.). These 'uncontrolled causes... are always strictly innumerable' (p. 18). The validity of tests of significance, Fisher (p. 19) argues, 'may be guaranteed against corruption by the causes of disturbance which have not been eliminated' by means of direct control. Howson and Urbach (1989, p. 147) claim that randomization applies only to a 'relatively small number of the trials that are actually conducted', as most trials are not aiming at significance tests, but parameter estimation. Their claim does not hold, because the estimates may also be biased due to the 'corrupting' effect of uncontrolled nuisance variables. Another objection to randomization as a catch-all solution to mis-specification is better founded (p. 149): it might occur that the impact of a nuisance variable is influenced by the experimental design. For example, consider a potato-growing trial, where each of two breeds are planted at two different spots, or are randomly planted on one larger spot. It might be that each of the varieties benefits likewise from the attention of a particular insect if planted apart, but that the insect has a preference for one of the breeds if they are planted randomly. Indeed, it might be. Any inference, however, will be affected by such a conceivable 'corrupting' factor. Randomization is a means of reducing such bias, it is not a guarantee that no bias will remain. Howson and Urbach (p. 145) maintain that 'the essential feature of a trial that permits a satisfactory conclusion as to the causal efficacy of a treatment is whether it has been properly controlled'. This is overly optimistic with regard to the feasibility of control. Moreover the more non-theoretical an empirical statistical problem is, the more randomization matters: control is pretty hard if one does not know what to control for.

Randomization also affects the economy of research. Rather than spending resources on eliminating as many of these causes as possible, the researcher should randomize. Randomization, rather than introducing assumptions about the relevant statistical distribution, provides the ('physical') basis of inductive inference. It is needed to cope with the innumerable causes that are merely a nuisance to the researcher.

Fisher ([1935] 1966, p. 102) adds that uniformity of experiments is not a virtue:

> The exact standardisation of experimental conditions, which is often thoughtlessly advocated as a panacea, always carries with it the real disadvantage that a highly standardised experiment supplies direct information only in respect of the narrow range of conditions achieved by standardisation.

It leads to idiosyncratic inferences. A similar remark can be made with regard to model design.

Fisher's theory of experimental design has little impact on most applied econometrics: economic data are usually non-experimental. Still, during the rise of econometric theory, experimental design played an important role in the econometric language (in particular Haavelmo's). The tension between this theory and econometric practice is the topic of chapter 6, section 4.

3.4 Fiducial inference

An important ingredient of Fisher's theory of inference is fiducial probability. In contrast to his other contributions, fiducial inference never gained much support, primarily because of its obscurity. It may be of interest for the interpretation of mainstream econometrics, however, because one may interpret arguments of econometricians as implicitly based on a fiducial argument.[11]

Consider a 95% confidence interval for a random variable X which is normally distributed as $X \sim N(\theta, 1)$:[12]

$$X - 1.96 < \theta < X + 1.96. \tag{4}$$

The meaning of this confidence interval is that the random interval $\langle X - 1.96, X + 1.96\rangle$ contains θ with a probability of 95%. In Fisher's view, θ is not random, but X is. This leads to a problem of interpretation, if a particular value x of X is observed, say $x = 2.14$. Can we plug this value in, retaining the interpretation of a confidence interval, i.e. is it justifiable to hold that in this case θ lies with 95% probability between 0.18 and 4.10? As long as θ is not considered a random quantity, the answer is no. You cannot deduce a probability from non-probabilistic propositions. The Bayesian solution is to regard θ as a random variable. For inference, a prior probability distribution is needed, which Fisher holds problematic in many (if not most) practical situations. Fisher's solution to the inference problem is to invoke a probabilistic *Gestalt switch*, the fiducial distribution $\phi(y)$. Since $X - \theta$ is distributed $N(0, 1)$, so is $\theta - X$. The fiducial distribution of θ is now defined as

$$P(\theta - X \leq y) = \phi_\theta(y) \forall y. \tag{5}$$

Given a fixed $X = x$, and granted that θ is not random, Fisher assigns this probability distribution to the non-random variable θ. But $P(\theta - X \leq y)$ is not the same as $P(\theta - x \leq y)$. There is no justification for the subtle change, and hence for the fiducial probability function $\phi_\theta(y)$.

Fisher considers his fiducial probability theory as a basis for inductive reasoning if prior information is absent (see Fisher, 1956, p. 56, where he also notes, that, in earlier writings he considered fiducial probability as a substitute for, rather than a complement to, Bayesian inference). Fiducial inference remains closer to the spirit of Bayesian inference than the Neyman–Pearson methodology, discussed below. Fisher (p. 51) remarks that the concept of fiducial probability is 'entirely identical' with Bayesian inference, except that the method is new. Blaug (1980, p. 23), on the other hand, writes that fiducial inference is 'virtually identical to the modern Neyman–Pearson theory of hypothesis testing'. This is a plain misinterpretation.[13] Neyman (1952, p. 231) argues that there is no relationship between the fiducial argument and the theory of confidence intervals. Neyman (1977, p. 100) is more blunt: fiducial inference is 'a conglomeration of mutually inconsistent assertions, not a mathematical theory'.

The problem of a fiducial probability distribution is that it makes sense only to those who perceive Fisher's *Gestalt switch*, and they are few. Apart from this, there are numerous technical problems in extending the method to more general problems than the one considered. Fiducial inference is at best controversial and hardly, if ever, invoked in econometric reasoning. Summarizing, Fisher's theory of inference is not a strong basis for econometrics, although a number of the more technical concepts he introduced are very important and useful in other interpretations of probability.

4 The Neyman–Pearson approach to statistical inference

In a series of famous papers, the Polish statistician Jerzy Neyman and his English colleague Egon Pearson developed an operational methodology of statistical inference (Neyman and Pearson, 1928, 1933a, b). It is intended as an alternative to inverse inference, and, initially, also as an elaboration of Fisher's methods (although Fisher was not amused). Neyman and Pearson invented a frequentist theory of testing, where the aim is to minimize the relative frequency of errors.

4.1 Inductive behaviour

Neyman and Pearson consider their theory as one of inductive behaviour, instead of a theory of inductive inference. Behaviourism (created by Ivan Pavlov, and, in particular, by John B. Watson) was the dominant school of thought in psychology from the 1920s till the 1940s. Neyman and Pearson (1933a, pp. 141–2) follow suit and

> are inclined to think that as far as a particular hypothesis is concerned, no test based upon the theory of probability can by itself provide any valuable evidence of the truth or falsehood of that hypothesis.
>
> But we may look at the purpose of a test from another view-point. Without hoping to know whether each separate hypothesis is true or false, we may search for rules to govern our behaviour with regard to them, in following which we insure that, in the long run of experience, we shall not be too often wrong.

Neyman and Pearson hold inference as a decision problem where the researcher has to cope with two possible risks: rejecting a true null hypothesis, a so-called Type I error, and accepting a false one, a Type II error (Neyman and Pearson, 1928). Only in exceptional cases is it possible to minimize both types of error simultaneously.

In a classic example, Neyman and Pearson (1933a, p. 146) wonder whether it is more damaging to convict an innocent man or to acquit a guilty. Mathematical theory may help in showing how to trade off the risks. Usually, the Type I error is considered to be the more serious one (although this is certainly not always the case: testing for unit roots is an example where in the frequency approach the Type II error will be more dangerous than the Type I error). In practice, the probability of a Type I error has a preset level, usually Fisher's 5% or 1%. This is the *significance level*, α. Given α, the goal is to find a test that is most *powerful*, that is one which minimizes the probability of a Type II error. An optimum test is a *uniformly most powerful* (UMP) test, given significance level α.

The Neyman–Pearson methodology is the first one in the statistical literature which takes explicit account of the alternatives against which a hypothesis is tested. Without specifying the alternative hypothesis, it is not possible to study the statistical properties (in particular, the power) of a test. In that case, it is unknown when a test is optimal (see Neyman,

1952, pp. 45 and 57). This seems to be an important step ahead, but Fisher (1956, p. 42) is not impressed:

> To a practical man, also, who rejects a hypothesis, it is, of course, a matter of indifference with what probability he might be led to accept the hypothesis falsely, for in his case he is not accepting it.

Is the indifference of Fisher's practical man justified?

4.2 The Neyman-Pearson lemma

The goal of Neyman and Pearson (1933a) is to find 'efficient' tests of hypotheses based upon the theory of probability, and to clarify what is meant by this efficiency. In 1928 they conjectured that a likelihood ratio test would provide the criterion appropriate for testing a hypothesis (Neyman and Pearson, 1928). This paper shows that the likelihood ratio principle serves to derive a number of known tests, such as the χ^2 and the *t*-test. The Neyman–Pearson (1933a) lemma proves that these tests have certain optimal properties.

The efficiency of a testing procedure is formulated in the properties of the critical region. Alternative tests distinguish themselves by providing different critical regions. An efficient test picks out the 'best critical region', i.e. the region where, given the Type I error, the Type II error is minimized. If more test procedures satisfy this criterion for efficiency, then the UMP test should be selected (if available).

If H_0 is the null hypothesis, and H_1 the alternative hypothesis, then (in Neyman and Pearson's 1933a, p. 142, terminology) H_0 and H_1 define the populations of which an observed sample $x = x_1, x_2, \ldots x_n$ has been drawn. Further, $p_0(x)$ is the probability given H_0 of the occurrence of an event with x as its coordinates (a currently more familiar notation is $P(x|\theta)$, where θ denotes the parameterization under H_0). The probability that the event or sample point falls in a particular critical region ω is (in the continuous case):

$$P_0(\omega) = \int_\omega \cdots \int p_0(x)dx. \qquad (6)$$

The optimum test is defined as the critical region ω satisfying

$$P_0(\omega) \leq \alpha, \qquad (7)$$

and

$$P_1(\omega) = \text{maximum}. \tag{8}$$

Neyman and Pearson (1933a) prove that such an optimum test exists in the case of testing simple hypotheses (testing $\theta = \theta_0$ against $\theta = \theta_1$) and is provided by the likelihood ratio statistic λ:

$$\lambda = \frac{P_0(x)}{P_1(x)}. \tag{9}$$

The Neyman–Pearson lemma states that the likelihood ratio statistic provides an optimal critical region, i.e. given a significance level α, it optimizes the power of the test. The theory of Fisher does not provide an argument for selecting an optimal test. In this sense, Neyman and Pearson provide an important improvement. Moreover, they present an extra argument for the application of maximum likelihood, and show that sufficient statistics contain all relevant information for testing (simple) hypotheses. Neyman and Pearson note that testing composite hypotheses is slightly more complicated. For testing composite hypotheses (e.g. $\theta = \theta_0$ against $\theta > \theta_0$), a uniformly most powerful test (one that maximizes power for all alternatives $\theta_1 > \theta_0$) exists if the likelihood ratio is monotonic (one-parameter exponential distribution families belong to this class, see Lehmann, [1959] 1986, p. 78). In case of multi-parameter problems, a UMP test typically does not exist. By imposing additional conditions, one may find UMP tests within this restricted class of tests. An example is the class of unbiased tests (Mood, Graybill and Boes, 1974, p. 425).

4.3 Fisher's critique

The disagreement between Fisher and Neyman–Pearson is not just one of different kinds of mathematical abstraction. Indeed, Fisher holds that his theory is the appropriate one for a free scientific society, whereas Neyman–Pearson's either belongs to the world of commerce or to that of dictatorship. In a critique of the behaviouristic school (of Neyman, Pearson, Wald and others), Fisher (1956, p. 7; see also pp. 100–2) writes:

> To one brought up in the free intellectual atmosphere of an earlier time there is something rather horrifying in the ideological movement represented by the doctrine that reasoning, properly speaking, cannot be applied to lead to inferences valid in the real world.

Fisher (1955, p. 77) opposes the use of loss functions for the purpose of scientific inference:

in inductive inference we introduce no cost functions for faulty judgements, for it is recognized in scientific research that the attainment of, or failure to attain to, a particular scientific advance this year rather than later, has consequences, both to the research programme, and to advantageous applications of scientific knowledge, which cannot be foreseen. In fact, scientific research is not geared to maximize the profits of any particular organization, but is rather an attempt to improve *public* knowledge.

In his methodological testament, Fisher (1956, p. 88) concludes that the principles of Neyman and Pearson will lead to 'much wasted effort and disappointment'.

Is the behaviouristic interpretation appropriate for econometrics? Many empirical papers in the economic journals do not deal explicitly with the behaviour of the researcher or policy maker whom he advises. The risks of making a wrong decision are not further evaluated, their monetary or utilitaristic implications rarely discussed. Even in cases where risks are appropriately evaluated, the Neyman–Pearson approach is not always the optimal one (Berger, [1980] 1985).

Less persuasive is Fisher's argument against the relevance of power and its trade-off with the size of a test. His opposition may have been partly emotional, as Fisher used a similar but mathematically less well-defined notion: sensitivity (see Fisher, [1925] 1973, p. 11). Neyman and Pearson were quite right to investigate the efficiency of alternative statistical tests. If decision making is hampered by conflicting test results, caused by using a number of alternative but equally valid tests, the selection of the most efficient test avoids the fate of Buridan's ass. On the other hand, Fisher's motivation for his opposition, i.e. that alternative hypotheses are not always well defined (and certainly not restricted to specific parametric classes such as in Neyman–Pearson methods) holds in many practical situations. Fisher would prefer robust tests rather than UMP tests.

An inevitable weakness of the Neyman–Pearson theory (acknowledged by its inventors) is the arbitrariness of the criterion for choosing or rejecting a hypothesis (the same weakness holds for Fisher's methodology). If the convention of applying a significance level of 0.01 or 0.05 is followed, then Lindley's 'paradox' shows that with growing sample size, any hypothesis will be rejected.[14] Econometricians tend to ignore the fact that the significance level should vary with sample size: macro-econometric studies using a time series of observations spanning 30 years and micro-econometric studies reporting statistics of cross sections consisting of 14,000 observations often report tests using the same 0.01 or 0.05 significance levels (see the evidence in Keuzenkamp and Magnus, 1995).

The Neyman–Pearson methodology has been criticized by McCloskey (1985) as an example of statistical rhetoric. Fisher (1956, pp. 99–100) had already made a similar point:

An acceptance procedure is devised for a whole class of cases. No particular thought is given to each case as it arises, nor is the tester's capacity for learning exercised. A test of significance on the other hand is intended to aid the process of learning by observational experience. In what it has to teach each case is unique, though we may judge that our information needs supplementing by further observations of the same, or of a different kind. To regard the test as one of a series is artificial; the examples given have shown how far this unrealistic attitude is capable of deflecting attention from the vital matter of the weight of the evidence actually supplied by the observations on some possible theoretical view, to, what is really irrelevant, the frequency of events in an endless series of repeated trials which will never take place.

Finally, many followers of the Neyman–Pearson methodology pay lip service to its principles, without explicating the losses that will result from decisions, and without the adaptation of Neyman and Pearson's framework of repeated sampling.

4.4 Neyman–Pearson and philosophy of science

Ronald Giere (1983) uses a decision theoretic framework very similar to the Neyman–Pearson methodology (without ever mentioning them) for a theory of testing theoretical hypotheses in a hypothetico-deductive method. But Giere acknowledges that science does not always necessitate making practical decisions, in particular if the evidence is not conclusive. A decision-making approach for testing scientific theories should, therefore, be accompanied by a theory of satisficing *á la* Simon. This satisficing approach enables the scientist to withhold judgement, if the risk of error is valued more highly than the necessity to make a decision. The 'institution of science' decrees the satisfaction levels (they may differ in different disciplines). Although Giere's approach has the merit of being non-dogmatic, it lacks clear guidelines.

Not only Giere has tried to encapsulate the Neyman–Pearson methodology. In one of his rare remarks on statistics, Lakatos (1978, p. 25) praises the merits of the Neyman–Pearson methodology. He relates it to the conjecturalist strategy in the philosophy of science:

Indeed, this methodological falsificationism is the philosophical basis of some of the most interesting developments in modern statistics. The Neyman–Pearson approach rests completely on methodological falsificationism.

Note, however, that the Neyman–Pearson approach to statistics has the purpose of guiding behaviour, not inference. It also predates Popper's *Logik der Forschung* (the first German edition appeared in 1934). Hence, Lakatos is wrong. Moreover, Popper ([1963] 1989, p. 62) explicitly rejects behaviourism:

> Connected with, and closely parallel to, operationalism is the doctrine of *behaviourism*, i.e. the doctrine that, since all test-statements describe behaviour, our theories too must be stated in terms of possible behaviour. But the inference is as invalid as the phenomenalist doctrine which asserts that since all test-statements are observable, theories too must be stated in terms of possible observations. All these doctrines are forms of the verifiability theory of meaning; that is to say, of induction.

It is, therefore, no surprise that Popper does not refer to the Neyman–Pearson methodology as a statistical version of his methodology of falsificationism. His reluctance to even mention Neyman and Pearson is not 'one of those unsolved mysteries in the history of ideas' as Blaug (1980, p. 23) asserts, but is the consequence of Popper's strict adherence to methodological falsificationism.

Neyman–Pearson methodology in practice, finally, is vulnerable to data-mining, an abuse on which the different Neyman–Pearson papers do not comment. It is implicitly assumed that the hypotheses are *a priori* given, but in many applications (certainly in the social sciences and econometrics) this assumption is not tenable. The power of a test is not uniquely determined once the test is formulated after an analysis of the data. Instead of falsificationism, this gives way to a verificationist approach disguised as Neyman–Pearson methodology. I will now turn to Popper's own ideas concerning probability.

5 Popper on probability

A good account of probability is essential to the practical relevance of methodological fallibilism (see Popper, [1935] 1968, p. 146, cited above in chapter 1, section 5.1). Has Popper been able to meet this challenge? This section deals with Popper's theories of probability in order to answer the posed question. His theories deal with von Mises' frequency theory in conjunction with a critique of epistemological probability. Later, he changed his mind and formulated a 'propensity' theory which should enable analysis of the probability of single events (instead of collectives). Another change of mind was the formulation of a corroboration theory, which shares some of the intentions of an epistemological theory but suffers from much greater weaknesses.

5.1 Popper and subjective probability

Popper ([1935] 1968, p. 146) distinguishes two kinds of probability: related to events and related to hypotheses. These may be interpreted as $P(e|h)$ and $P(h|e)$ respectively. The first is discussed in a framework similar to von Mises' frequency theory, the second is epistemological, and belongs to the realm of corroboration theory. Popper aims to repair the defects of von Mises' theory,[15] and criticizes efforts (by, among others, Reichenbach) to translate a probability of a hypothesis into a statement about the probability of events (p. 257). A reflection on Bayes' theorem shows that the radical proposition of Popper is untenable. Still, a brief discussion of Popper's views on the two separate kinds of probability is useful for an appraisal of his philosophy of science.

Despite his own verdict, Popper's contribution to von Mises' theory is, at most, marginal. Indirectly, however, he may have triggered the solution to one of the formal weakness of von Mises' theory. In 1935, he discussed, at Karl Menger's mathematical colloquium in Vienna, the conceptual problem of the collective. Wald was one of the delegates (other participants included Gödel and Tarski) and, after hearing Popper, he became interested in the problem. This resulted in the famous paper in which Wald 'repaired' the definition of collectives.

This positive contribution to the frequency theory of probability is accompanied by a number of negative statements about the epistemological theory of probability, in particular the logical theory of Keynes (discussed in the next chapter). It is true that this theory is not without problems, but Popper's arguments against it are not strong. Two examples, taken from Popper (1983), may give an impression of Popper's argumentation.

The first example (p. 294) deals with Kepler's first law, which can be summarized by the statements:
(a) x is an ellipse;
(b) x is a planetary orbit;
and the inference: $P(a|b) = 1$.[16] This is the same as to say that all planetary orbits will be ellipses. Now Popper identifies this statement with an observed (empirical) frequency, which he calls a probability. On the other hand, the *a priori* probability (or 'logical frequency') of such a statement 'must' be equal to zero given the infinity of alternative logical possibilities (see also Popper, [1935] 1968, p. 373). This zero *a priori* probability of the conjecture conflicts with the 'true' frequentist probability which is nearly equal to one. Because of the apparent contradiction, Popper rejects the logical probability propositions.

But are prior probabilities necessarily equal to zero? This is an important question indeed. In the next two chapters it is argued that Popper is wrong. First, a probability is not just a number, but like time and heat can only be measured with respect to a well-specified reference class. A molecule is big compared with a quark, but may be small relative to Popper's brain. Similarly, a particular theory is not probable in itself, but more (or less) probable than a rival theory. Secondly, even given a large number of conceivable rival theories in the reference class, there are good arguments to assign different prior probability values to those members, in relation to their respective simplicity (see chapter 5).

Continuing the example, an investigator (who even ignores the simplicity ranking) may believe that it is *a priori* equally probable that orbits are squares or circles. This would assign a prior probability of one half to the square theory versus one half for the circle theory. Combining these prior probabilities with data will result in an *a posteriori* probability, and it is no bold conjecture that the circle theory will out-perform the square theory rather quickly. A next step, to improve the fit of a theory with the observations, is to introduce a refinement, say the ellipsoidal theory. Again, the data will dominate the prior probabilities quickly. This is how, in a simplistic way, probability theory can be applied to the appraisal of theories without paradox.

If, however, Popper's claim is accepted that the number of possible and presently relevant conjectures is infinite and all are equally likely, then another puzzle should be solved: how is it possible that scientists come up with conjectures that are not too much at odds with the observed data? Guessing is a dumb man's job, as Feynman said: inference must inevitably be based on inductive evidence.

Consider a second example of Popper discussing probability. Popper (1983, pp. 303–5) introduces a game, to be played between a frequentist Popperian and an inductivist. It is called Red and Blue, and the Popperian will almost certainly win. The idea is that you toss a coin, win a dollar if heads turns up, or lose a dollar with tails. Now repeat the game indefinitely, and look at the total gains. If they are zero or positive, they are called blue, red if negative. At the first two throws, the player loses and from then on plays even. The game is repeated for a total of 10,000 throws, with the last throw leading exactly to the borderline between red and blue. Now, according to Popper, the inductivist would gamble ten thousand to one that the next throw would keep him in the red zone, while the keen Popperian frequentist would gratefully accept this offer (or any offer with odds higher than one to one).

It is obvious that this game is flawed and surprising that Popper dares to use it as an argument. Implicitly, Popper gives the frequentist knowledge of the structure of the game, whereas this information is hidden for the inductivist. The argument would collapse once information is distributed evenly. It cannot be used as an argument against inductivism, except if Popper would agree that, in scientific inference, the frequentist knows the laws of nature beforehand. The problem in science, of course, is that these laws are unknown and have to be inferred from data.

Other objections of Popper to the subjective probability theory can easily be dismissed. As a final example consider Popper's 'paradox' that, according to the subjective view, tossing a coin may start with an *a priori* probability of heads of 1/2, and after 1,000,000 tosses of which 500,000 are heads, ends with an *a posteriori* probability of 1/2 as well, which 'implies' that the empirical evidence is irrelevant information (Popper, [1935] 1968, p. 408). This means, according to Popper, that a degree of reasonable belief is completely unaffected by accumulating evidence (the weight of evidence), which would be absurd. It is, of course, a malicious interpretation of subjective probability to identify a posterior distribution with a posterior mean.

5.2 The propensity theory of probability

After writing the *Logic of Scientific Discovery*, Popper succumbed to the temptation of von Mises' forbidden apple, the probability of single events. Popper (1982, pp. 68–74) is interested in specific events related to the decay of small particles in quantum theory. The propensity theory, designed for this purpose, is an extension of the indifference theory, with a distant flavour of the frequency theory. A propensity is a (virtual) relative frequency which is a physical characteristic of some entity (p. 70). A propensity is also interpreted as an (objective) tendency to generate a relative frequency (p. 71). As such, it is an *a priori* concept, although Popper argues that it is testable by comparing it to a relative frequency. But then the 'collective' enters again, and it is unclear how the outcome of one single event (for example, throwing heads) can be related to the long-run frequency of events.

The propensity theory has not been very successful for the analysis of small particle physics (see Howson and Urbach, 1989, pp. 223–5). Its relevance for the social sciences, in particular econometrics, is small, except if people's desires are based on inalienable physical characteristics. Popper did not claim that the theory should be of interest for the social sciences, and in fact, he was not interested in discussing his views on probability with the probability theorists and statisticians at LSE.[17] Hunger may be a physical stimulus to eating, but the choice between bread or rice cannot be explained in such terms. If propensities are understood as probabilistic causes, then they do not make sense as inverse probabilities, because that conflicts with notions of causality.

Summarizing, Popper's views on probability add few useful insights to existing 'objective' theories of probability. Reading his sequence of changing interpretations of probability is an irritation. If methodological falsificationism must have probabilistic underpinnings (as Popper acknowledges), then the best way forward is to accept either Fisher's theory of inference, or the theory of von Mises as a substitute for Popper's own views. But both acknowledge support of the inductivist arguments, which is the ultimate sin of Popper's philosophy of science.

5.3 Corroboration and verisimilitude

A methodological theory that pretends to separate the wheat from the chaff needs a method of theory appraisal. According to Popper, probability theory will not do. Falsifications count, not verifications. Probability theory (so it is asserted) cannot take care of this asymmetry. This raises two questions: first, what to do if two theories are as yet both unfalsified, and second, what to do if both are falsified. Popper's answer to the first question is to invoke the degree of corroboration, a hybrid of confirmation and content, discussed in Popper ([1935] 1968). The second question is the source of Popper's later ideas on closeness to the truth, verisimilitude.

A theory is said to be corroborated as long as it stands up against tests. Some tests are more severe than others, hence there are also different degrees of corroboration. The severity of tests depends on the degree of testability, and that, in turn, is positively related to the simplicity of the theory. The simpler a theory, Popper argues, the lower its *a priori* (logical) probability (this is in sharp conflict with the ideas presented in chapter 5). The degree of corroboration of a theory which has passed severe tests is inversely related to the logical probability (p. 270), but is positively related to a degree of confirmation (the two make a hybrid which leads at times to a muddle in Popper's argument, and also has led to unnecessary arguments with Carnap).

Popper denotes the degree of confirmation of hypothesis h by evidence e by $C(h,e)$. What is the relation to the probability $P(h|e)$? Popper (pp. 396–7) argues that $C(h, e) \neq P(h|e)$. I will not interfere with his argument, but discuss how Popper wants to measure $C(h, e)$. In a footnote to the main text of his book, (p. 263), he acknowledges that

It is conceivable that for estimating degrees of corroboration, one might find a formal system showing some limited formal analogies with the calculus of probability (e.g. with Bayes' theorem), without however having anything in common with the frequency theory [of Reichenbach].

The (later written) appendix of the *Logic* proceeds in this direction by endorsing (with a few addenda) an information-theoretical definition of corroboration (information theory dates back to the work of Claude Shannon and Norbert Wiener in 1948; see Maasoumi, 1987, for discussion and references):

$$C(h, e) = \log_2\left(\frac{P(e|h)}{P(e)}\right) \tag{10}$$

(where $\log_2$ denotes logarithm with base 2). A good test is one that makes $P(e)$ small, and $P(e|h)$ (Fisher's likelihood) large. Popper's discussion is problematic, as he interprets e as a statistical test report stating a goodness of fit. This cannot be independent of the hypothesis in question, however. Popper ([1935] 1968, p. 413) finally links his measure of corroboration directly to the likelihood function. After much effort, we are nearly back to the dilemmas of Fisher's theory. The essential distinction with Fisher is Popper's (p. 418) emphasis that '$C(h, e)$ can be interpreted as a degree of confirmation only if e is a report on the severest tests we have been able to design'. That is, Popper invokes the goal of tests for the interpretation of the results, to exploit the asymmetry between falsification and verification. If two theories are falsified, then additional criteria are needed, which may be found in the theory of verisimilitude.

A confirmation of a highly improbable theory, one which implies a low $P(e)$, adds much to the degree of corroboration. That appeals to common sense (although Popper seems to confuse $P(e)$ and $P(h)$ in his discussion of 'content', e.g. p. 411). But should we, therefore, do as much as possible to formulate theories with as low as possible *a priori* logical probabilities? Is, in the words of Levi (1967, p. 111), the purpose of inquiry to nurse our doubt and to keep it warm? The most sympathetic way to interpret Popper is to regard his degree of corroboration as the gain of information, which a test may bring about. Then the goal of scientific inference is not to search for *a priori* highly improbable theories, but to gain as much information as possible. Counting another raven to corroborate the hypothesis that all ravens are black does not add much information and is, for sake of scientific progress, not very helpful and a rather costly way to corroborate the theory that all ravens are black. Systematically confirming a prediction of a highly disputed theory, on the other hand, is worth its effort.

In Popper ([1963] 1989, chapter 10), a new concept is added to his theory of scientific inference in order to discuss matters of progress in science. Most scientists agree that empirical theories can never claim to have reached a perfect match with 'truth'. But we do know that Newton's theory provides a better approximation to observable facts than Keppler's. Popper's new idea is the concept of the *verisimilitude* of a theory, a combination of closeness to the truth (empirical accuracy) and the content of a theory. The definition Popper gives seems appealing (p. 233): a theory is closer to the truth than another one, if the first has more true, and fewer false, consequences than the latter. This criterion satisfies Lakatos' demand for a 'whiff of inductivism', to study scientific progress as a sequence of improvements rather than instant crucial tests.

However, the definition of verisimilitude is untenable because false theories have equal verisimilitudes (see for discussion Watkins, 1984). A false theory (one with at least one false consequence) can never be closer to the truth than another false one. As no empirical theory (model) is perfectly flawless, this leads to the extreme scepticism of which Russell accuses the early Greek sceptics. Kuhn (1962, pp. 146–7) was right when he argued that

> if only severe failure to fit justifies theory rejection, then the Popperians will require some criterion of 'improbability' or of 'degree of falsification.' In developing one they will almost certainly encounter the same network of difficulties that has haunted the advocates of the various probabilistic verification theories.

Popper's subsequent theories of probability are, in short, of little help for completing methodological falsificationism. His critique of other theories of probability is of little interest: weaknesses which are real have been more effectively detected, analysed and sometimes repaired by proper probability theorists.

6 Summary

How useful for econometric inference are the different interpretations of probability given so far? The (early formulation of the) indifference theory is clearly unsatisfactory as a foundation for scientific inference in general and econometric inference in particular. Its relative, the propensity theory, is also of little use for econometrics. In contrast, the frequency interpretations presented in this chapter are widely accepted by econometricians. They are not bothered by the circularity of the frequency definition of probability.

Mainstream econometric methodology (as defined by the core chapters of textbooks like Chow, 1983; Johnston, [1963] 1984; Maddala, 1977; Hill, Griffith and Judge, 1997; Goldberger, 1998) is a blend of von Mises', Fisher's and Neyman–Pearson's ideas. This blend embraces the frequency concept of von Mises, his emphasis on objectivity, and asymptotic arguments. The conditions that warrant application of von Mises' theory (convergence and randomness) are rarely taken seriously, however. The blend further relies heavily on Fisher's box of tools, and sometimes takes his hypothetical sample space seriously. A careful consideration of Fisher's essential concept of experimental design is rarely made. The Neyman–Pearson legacy is the explicit formulation of alternative hypotheses – although the practical relevance of the power of tests is small.

For the purpose of scientific inference in the context of non-experimental data, this blend provides weak foundations. The Neyman–Pearson approach is not attractive, as decision making in the context of repeated sampling is not even metaphorically acceptable. The frequentist highway to truth, paved with facts only, is still in need of a bridge. Not even a bridge to probable knowledge can be given though. Fisher's fiducial bridge has a construction failure, and is rightly neglected in econometric textbooks and journals alike. Still, many empirical results that correspond to good common sense are obtained using the methods of Fisher and, perhaps, Neyman–Pearson.

Why are these methods so popular, if they lack sound justification? A pragmatic explanation probably is the best one. Fisher's maximum likelihood works like a 'jackknife', Efron (1986, p. 1) argues:

> the working statistician can apply maximum likelihood in an automatic fashion, with little chance (in experienced hands) of going far wrong and considerable chance of providing a nearly optimal inference. In short, he does not have to think a lot about the specific situation in order to get on toward its solution.

Similarly, one of the strongest critics of Fisher's approach, Jeffreys ([1939] 1961, p. 393), disagrees with the fundamental issues but states, reassuringly, that there is rarely a difference on actual inferences made, when an identical problem is at stake. Weak foundations do not prevent reasonable inference.

Alea jacta est. But economic agents do things other than throwing dice. Perhaps the appropriate method of inference, although probabilistic, will be different in kind from the view based on observed frequencies such as throws of dice. Let us, therefore, cross our Rubicon, and move on to the bank of epistemological probability.

Notes

1. Maddala (1977, p. 13) claims: 'The earliest definition of probability was in terms of a long-run relative frequency.' This claim is wrong: the relative frequency theory of probability emerged in response to the indifference theory.
2. The weak law relates to convergence in probability. Mood, Graybill and Boes (1974, p. 232) and von Mises (1964, pp. 230–43) provide discussions of the laws of large numbers of Khintchine and Markov which do not rely on the restriction on the variance (see also Spanos, 1986, chapter 9). The strong law relates to almost sure convergence.
3. Keynes ([1921] CW VIII, pp. 394–9) cites an amusing sequence of efforts to 'test' the law of large numbers. In 1849, the Swiss astronomer Rudolf Wolf counted the results of 100,000 tosses of dice, to conclude that they were ill made, i.e. biased (Jaynes, 1979, calculates the physical shape of Wolf's dice, using his data). Keynes [1921] CW VIII, p. 394) remarks that G. Udny Yule also 'indulged' in this idle curiosity. Despite Keynes' mockery, Yule ([1911] 1929, pp. 257–8; the ninth edition of the book to which Keynes refers) contains the statement: 'Experiments in coin tossing, dice throwing, and so forth have been carried out by various persons in order to obtain experimental verification of these results . . . the student is strongly recommended to carry out a few series of such experiments personally, in order to acquire confidence in the use of the theory.'
4. The expression of the second law strongly resembles von Mises' ([1928] 1981, p. 105) formulation of Bernoulli's theorem, which says: '*If an experiment, whose results are simple alternatives with the probability p for the positive results, is repeated n times, and if* ϵ *is an arbitrary small number, the probability that the number of positive results will be not smaller than* $n(p - \epsilon)$*, and not larger than* $n(p + \epsilon)$*, tends to 1 as n needs to infinity.*'
5. For any sequence, there is an admissible ϕ that coincides with a ϕ that is inadmissible, where admissibility is defined according to the definition stated in the randomness axiom.
6. A remarkable example is his research in human serology. In 1943, Fisher and his research group made exciting discoveries with the newly found Rh blood groups in terms of three linked genes, each with alleles. They predicted the discovery of two new antibodies and an eighth allele. These predictions were soon confirmed.
7. This phrase is addressed to Wald and Neyman. Von Mises does not appear in Fisher's writings, not even in Fisher's (1956), his most philosophical work.
8. Spanos (1989, p. 411) claims that this 'axiom' of correct specification was first put forward by Fisher. Spanos also asserts that Tjalling Koopmans (1937) refers to it as 'Fisher's axiom of correct specification'. Although Koopmans does discuss specification problems in his monograph, there is no reference to such an axiom. Indeed, it would have been surprising if Fisher had used the term 'axiom' in relation to correct specification: his

style of writing is far remote from the formalistic and axiomatic language of the structuralistic tradition. Fisher's hero was Darwin (or Mendel), not Hilbert (or Kolmogorov). He was interested, for example, in pig breeding, not axiom breeding (see the delightful biography by his daughter, Joan Fisher Box). His scepticism towards axiomatics is revealed in letters to H. Fairfield Smith and Harold Jeffreys, and an anecdote, all discussed by Barnard (1992, p. 8). Around 1950, after a lecture, Fisher was asked whether fiducial probability satisfied Kolmogorov's axioms. He replied: 'What are they?' The occasion where Fisher makes use of the notion 'axiom' is in ([1935] 1966, pp. 6–7) where he criticizes Bayes' 'axiom' (of a uniform prior probability distribution).

9. The definition in Fisher (1922, p. 309) is somewhat different: 'A statistic satisfies the criterion of consistency, if, when it is calculated from the whole population, it is equal to the required parameter.'
10. Fisher Box (1978, p. 147) quotes a classic 1926 paper of Fisher ('The arrangement of field experiments'), which states that after randomization, [t]he estimate of error is valid because, if we imagine a large number of different results obtained by different random arrangements, the ratio of the real to the estimated error, calculated afresh for each of these arrangements, will be actually distributed in the theoretical distribution by which the significance of the result is tested.'
11. Explicit references by econometricians to fiducial inference are rare. Koopmans (1937, p. 64) is an exception. He discusses exhaustive statistics and compares this to Neyman's confidence intervals. He does not pursue the issue more deeply and concludes that 'maximum likelihood estimation is here adopted simply because it seems to lead to useful statistics' (p. 65).
12. The following discussion is based on Lehmann ([1959] 1986, pp. 225–30).
13. Blaug is in good company. Statisticians of name, such as Bartlett, have argued that fiducial distributions are like distributions that provide Neyman-type confidence intervals (see Neyman, 1952, p. 230, for references).
14. For any level of significance α, and for any non-zero prior probability of a hypothesis ($P(h_0)$), there is always a sample size n such that the posterior probability of the hypothesis equals $1 - \alpha$. Thus, a hypothesis that has strong *a posteriori* support from a Bayesian perspective, may have very little support from a Fisherian point of view.
15. Popper met von Mises in Vienna (see Popper, 1976, chapter 20). Popper (1982, p. 68) claims that he solved the ambiguities in von Mises' frequency theory: 'I feel confident that I have succeeded in purging it of all those allegedly unsolved problems which some outstanding philosophers like William Kneale have seen in it.'
16. In 1734, Daniel Bernoulli won a prize at the French Academy with an essay on the problem of planetary motion. The problem discussed is whether the near coincidence of orbital planes of the planets is due to chance or not. Bernoulli's analysis is one of the first efforts to practise statistical hypothesis testing.

17. Denis Sargan (private discussion, 1 December 1992). See also De Marchi (1988, p. 33) who notes that Popper was not interested in discussions with LSE economists, as they smoked and Popper did not. Later on, Lakatos did try to get in touch with the LSE statisticians, but by then, according to Sargan, they had lost their appetite.

4 Probability and belief

Probability does not exist.

Bruno de Finetti (1974, p. x)

1 Introduction

Probability in relation to rational belief has its roots in the work of Bayes and Laplace. Laplace is known for his interest in the 'probability of causes', inference from event A to hypothesis H, $P(H|A)$. This is known as inverse probability, opposed to direct probability (inference from hypothesis to events, $P(A|H)$). Inverse probability is part of epistemology, the philosophical theory of knowledge and its validation. The acquisition of knowledge, and the formulation of beliefs, are studied in cognitive science. Of particular interest is the limitation of our cognitive faculties. In this chapter, the epistemological approaches to probability are discussed.

The unifying characteristic of the different interpretations of epistemological probability is the emphasis on applying Bayes' theorem in order to generate knowledge. All interpretations in this chapter are, therefore, known as Bayesian. Probability helps to generate knowledge, it is part of our cognition. It is not an intrinsic quality that exists independently of human thinking. This is the message of the quotation from de Finetti, featuring as the epigraph to this chapter: probability does not exist. Probability is not an objective or real entity, but a construct of our minds.

This chapter first introduces the theory of logical probability of Keynes (section 2) and Carnap (section 3). Section 4 deals with the personalistic ('subjective') interpretation of probability theory. The construction of prior probabilities, one of the great problems of epistemological probability, is the topic of section 5. Section 6 appraises the arguments.

2 Keynes and the logic of probability

The logical interpretation of probability considers probability as a 'degree of logical proximity'. It aims at assigning truth values other than zero or one to propositions for inference of partial entailment. The numbers zero and one figure as extreme cases. A probability of zero indicates impossibility, a probability equal to one indicates the truth of a proposition.

John Maynard Keynes belongs to the founders of this approach. The publication of *A Treatise on Probability* (Keynes, [1921] CW VIII) was an important event in the history of the theory of probability and induction (Russell, 1927, p. 280, writes that Keynes provides by far the best examination of induction known to him). Rudolf Carnap, Jaakko Hintikka and Richard Jeffrey continued research on logical probability.

2.1 A branch of logic

Keynes ([1921] CW VIII) regards probability theory, like economics, as a branch of logic. This logic is not the formal logic of Russell and Whitehead, but a more intuitive logic of practical conduct (see Carabelli, 1985). This theory of conduct is inspired by the Cambridge philosopher, G. E. Moore, and can be traced back to the writings of Hume.

The logical theory uses the word 'probability' primarily in relation to the truth of sentences, or propositions. The occurrence of events is of interest only insofar as the probability of the occurrence of an event can be related to the truth of the proposition that the event will occur.[1] A relative frequency is, therefore, not identical to a probability, although it may be useful in deriving a probability. Jeffreys ([1939] 1961, p. 401) provides the analogy that physicists once described atmospheric pressure in terms of millimetres, without making pressure a length. Probability is a rational degree of belief.

An important axiom of Keynes' system is his first one:

> Provided that a and h are propositions or disjunctions of propositions, and that h is not an inconsistent conjunction, there exists one and only one relation of probability P between a as conclusion and h as premiss. ([1921] CW VIII, p. 146)

This axiom expresses what Savage ([1954] 1972) calls the *necessary* view of probability, which he attributes to Keynes as well as to Jeffreys.[2] In this sense, Keynes' theory is an objective one, as he already makes clear in the beginning of his book (Keynes [1921] CW VIII, p. 4):

> in the sense important to logic, probability is not subjective. It is not, that is to say, subject to human caprice. A proposition is not probable because we think it so. When once the facts are given which determinate our knowledge, what is probable or improbable in these circumstances has been fixed objectively, and is independent of our opinion.

This view has been criticized by Ramsey, who defends a strictly subjective, or personalistic, interpretation of probability. Keynes ([1933] CW X, p. 339) yielded a bit to Ramsey.[3] Despite this minor concession, von Mises ([1928] 1981, p. 94) calls Keynes 'a persistent subjectivist'. This interpretation may be induced by Keynes' definition of probability as a degree of belief relative to a state of knowledge. However, persons with the same state of knowledge must have the same probability judgements, hence there is no room for entrenched subjectivity.

The most striking differences between Keynes and von Mises are:

- according to von Mises, the theory of probability belongs to the empirical sciences, based on limiting frequencies, while Keynes regards it as a branch of logic, based on degrees of rational belief; and
- von Mises' axioms are idealizations of empirical laws, Keynes' axioms follow from the intuition of logic.

It is remarkable that such differences in principles do not prevent the two authors from reaching nearly complete agreement on almost all of the mathematical theorems of probability, as well as the potentially successful fields of the application of statistics!

2.2 Analogy and the principle of limited independent variety

A central problem in the *Treatise on Probability* is to obtain a probability for a proposition (hypothesis) H, given some premise (concerning some event or evidential testimony), A. The posterior probability of H given A is given by Bayes' rule,

$$P(H|A) = \frac{P(A|H)P(H)}{P(A)}. \tag{1}$$

In order to obtain this posterior probability, a strictly positive prior probability for H is needed: if $P(H) = 0$, then $P(H|A) = 0 \; \forall A$. It is not obvious that such a prior does exist. If infinitely many propositions may represent the events, then without further prior information their logical *a priori* probability obtained from the 'principle of indifference' goes to zero. In philosophy, this is known as the problem of the zero probability of laws in infinite domain (Watkins, 1984).

Informative analogies are an important source for Keynesian probability *a priori*. Hume already noted that analogy is needed in inductive reasoning. Keynes ([1921] CW VIII, p. 247) cites Hume, who argues that reasoning is founded on two particulars: 'the constant conjunction of any two objects in all past experience, and the resemblance of a present object to any of them'. Keynes (p. 243) makes a crucial addition by introducing 'negative analogy'. Counterparts to negative analogy in contemporary econometrics are identification and, one might argue, global sensitivity analysis. Using the question 'are all eggs alike?' as an example, Keynes argues that the answer depends not only on tasting eggs under similar conditions.[4] Rather, the experiments should not be 'too uniform, and ought to have differed from one another as much as possible in all respects save that of the likeness of the eggs. [Hume] should have tried eggs in the town and in the country, in January and in June.' New observations (or replications) are valuable in so far they increase the variety among the 'non-essential characteristics of the instances'. The question then is whether in particular applications of inductive inference, the addition of new data adds simultaneously essential plus non-essential characteristics, or only non-essential characteristics. The latter provides a promising area for inductive inference. The statement is in line with another author's conclusion, that 'by deliberately varying in each case some of the conditions of the experiment, [we may] achieve a wider inductive basis for our conclusions'. (R. A. Fisher [1935] 1966, p. 102).

Keynes ([1921] CW VIII, p. 277) argues, that, if every separate configuration of the universe were subject to its own governing law, prediction and induction would become impossible. For example, consider the regression equation,

$$y_i = a_1 x_{i,1} + \ldots + a_k x_{i,k} + u_i, \quad (u_1, \ldots, u_n)' \sim (0, \sigma^2), \qquad (2)$$
$$i = 1, \ldots, n.$$

In this equation every observation i of y_i might have its own 'cause' or explanatory variable, x_j (k equals n). The same might apply to new observations, each one involving one or more additional explanatory variables. Alternatively, the parameters a might change with every observation, without recognizable or systematic pattern. These are examples of a lack of 'homogeneity' where negative analogy does not hold and, hence, new observations do not generate inductive insights.[5] The universe must have a certain amount of homogeneity in order to make inductive inference possible.

To justify induction, the fundamental logical premise of inference has to be satisfied. The *principle of limited independent variety* serves this purpose. It is introduced by Keynes in a chapter significantly entitled 'The justification of these [inductive] methods'.[6] Keynes (p. 279) defines the independent variety of a system of propositions as the 'ultimate constituents' of the system (the indefinable or primitive notions) together with the 'laws of necessary connection'.[7] The eggs example may be used to clarify what Keynes had in mind. The yolk, white and age of the egg are (perhaps not all) the ultimate constituents of the egg, while the chemical process of taste may be interpreted as a law of necessary connection. The chemical process is not different in space or time, and the number of ultimate constituents is low. Hence, independent variety seems limited.

If independent variety increases (for example by increasing the number of regressors, changes in functional form, etc., in equation 2), induction becomes problematic. Keynes argues that the propositions in the premise of an inductive argument should constitute a high degree of homogeneity:

> Now it is characteristic of a system, as distinguished from a collection of heterogeneous and independent facts or propositions, that the number of its premisses, or, in other words, the amount of independent variety in it, should be less than the number of its members.

Even so, a system may have finite or infinite independent variety. It is only with regard to finite systems that inductive inference is justified (p. 280). An object of inductive inference should not be infinitely complex, so 'complex that its qualities fall into an infinite number of independent groups', i.e. generators or causes (p. 287). If there is reason to believe that the condition of limited independent variety is met, then inductive inference is, in principle, possible. Keynes dubs this belief the *inductive hypothesis*. It is a sophisticated version of the principle of the uniformity of nature.[8] If independent variety is limited, the number of hypotheses h_i $(i = 1, \ldots, n)$ that may represent e is finite. Lacking other information, they are assigned a prior probability of $1/n$ in accordance with the principle of indifference (alternative measures are discussed in the next chapter).

Induction depends on the validity of the inductive hypothesis. The question is how to assess its validity. This cannot be done on purely logical grounds (as it pertains to an empirical matter), or on purely inductive grounds (this would involve a circular argument).[9] Keynes argues that there is no need for the hypothesis to be true. It suffices to attach a non-zero prior probability to it and to justify such a prior belief in each application. The domain of induction cannot be determined exactly by either logic or experience. But logic and experience may help to give at least some intuition about this domain: we know from experience that we may place some (though not perfect) confidence in the validity of limited independent variety in a number of instances.[10] 'To this extent', Keynes (pp. 290–1) argues, 'the popular opinion that induction depends upon experience for its validity is justified and does not involve a circular argument'. But one should be careful with applying probabilistic methods. In particular, Keynes' letter to C. P. Sanger discussing the problem of index numbers emphasizes this:

> when the subject is backward, the method of probability is more dangerous than that of approximation, and can, in any case, be applied only when we have information, the possession of which supposes some very efficient method, other than the probability method, available. ([1909] 1983 CW XI, p. 158)

As an example of such an alternative method, not yet available when Keynes wrote these comments, one might think of R. A. Fisher's theory of experimental design. Indeed, this is one of the few methods which has proved to be effective in cases of inference where the list of causes is likely to exceed the few captured in a regression model.

2.3 Measurable probability

Keynes has shown the possibility of obtaining logical prior probabilities, albeit in a very restricted class of cases. How much help is this for inference? A distinctive feature of Keynes' view is that not all probabilities are numerically measurable, and in many instances, they cannot even be ranked on an ordinal scale. The 'beauty contest', which has become famous due to the *General Theory* but had occurred previously, for another purpose, in the *Treatise*, is used to illustrate this point. Keynes ([1921] CW VIII, pp. 27–9) explains how one of the candidates of the contest sued the organizers of the *Daily Express* for not having had a reasonable opportunity to compete. Readers of the newspaper determined one part of the nomination. The final decision depended on an expert, who had to sample the top fifty of the ladies chosen by the readers.[11] The candidate complained in front of the Court of Justice, that she had not obtained an opportunity to make an appointment with this expert. Keynes argues that the chance of winning the contest could have been measured numerically, if only the response of the readers (who sent in their appraisals and thus provided an unambiguous ranking of the candidates) had mattered. The subjective taste of the single expert could not be evaluated in a similar way. Hence, a rational basis for evaluating the chances of the unfortunate lady was lacking. 'Keynesian' probability theory could not be used to estimate her loss.

The non-measurability of probabilities, the notion that very often probabilities cannot be assigned numerical values and often can, at best, be ranked on a cardinal scale, is what I call Keynes' incommensurability thesis of probability. This belongs to Keynes' most controversial contributions to probability theory. For example, Jeffreys ([1939] 1961, p. 16), who developed a probability theory similar to Keynes', proposes, as the very first axiom of his probability theory, that probabilities must be comparable. Indeed, the empirical applicability of probability theory would be drastically reduced if probabilities were incommensurable. A rational discussion about different degrees of belief would be impossible as soon as one participant claims that his probability statement cannot be compared with another one's (the resulting problem of rational discussion is similar to the utilitarian problem of rational redistribution given ordinal utility). One should be wary, therefore, of dropping measurable probability without proposing alternatives. More recently, statisticians like B. O. Koopman and I. J. Good, have tried to formalize probabilistic agnosticism. Ed Leamer may be regarded as a successor in econometrics (although he rarely refers to Keynes and Good, and never to Koopman).

2.4 Statistical inference and econometrics: Keynes vs Tinbergen

In the final part of the *Treatise*, Keynes discusses statistical inference. Whereas the logical theory is mainly directed towards the study of universal induction (concerning the probability that every instance of a generalization is true, e.g. the proposition that all swans are white), statistical inference deals with problems of correlation. This deals with the probability that any instance of this generalization is true ('most swans are white'; see Keynes, [1921] CW VIII, pp. 244–5). The latter is more intricate than, but not fundamentally different from, universal induction, Keynes argues (p. 447). The goal of statistical inference is to establish the analogy shown in a series of instances. Keynes (p. 454) comes fairly close to von Mises' condition of randomness, for he demands convergence to stable frequencies from given sub-sequences. Von Mises, however, is more explicit about the requirements for this condition.

Among econometricians, Keynes is known as one of their first opponents. This is due to his clash with Tinbergen in 1939. But Keynes had been tempted by the statistical analysis of economic data himself when he was working on *A Treatise on Probability* and employed at the London India Office. In his first major article, published in 1909 in the *Economic Journal* ('Recent economic events in India', 1983, CW XI, pp. 1–22), Keynes makes an effort to test the quantity theory of money by comparing estimates of the general index number of prices with the index number of total currency. The movements were surprisingly similar. In a letter to his friend Duncan Grant in 1908, Keynes writes that his 'statistics of verification' threw him into a

> tremendous state of excitement. Here are my theories–will the statistics bear them out? Nothing except copulation is so enthralling and the figures are coming out so

much better than anyone could possibly have expected that everybody will believe that I have cooked them. (cited in Skidelsky, 1983, p. 220)

Keynes himself remained suspicious of unjustified applications of statistics to economic data. A prerequisite of fruitful application of statistical methods is that positive knowledge of the statistical material under consideration must be available, plus knowledge of the validity of the principle of limited variety, knowledge of analogies and knowledge of prior probabilities. These circumstances are rare. Keynes ([1921] CW VIII, p. 419) argues, that

To apply these methods to material, unanalysed in respect of the circumstances of its origin, and without reference to our general body of knowledge, merely on the basis of arithmetic and of those of the characteristics of our material with which the methods of descriptive statistics are competent to deal, can only lead to error and to delusion.

They are the children of loose thinking, the parents of charlatanism. This provides the basis for Keynes' ([1939] CW XIV) later objections to the work of Tinbergen, to which I will turn now.

Tinbergen had made a statistical analysis of business cycles at the request of the League of Nations. Tinbergen's (1939a) major justification for relying on statistical analysis is that it may yield useful results, and may be helpful in measuring the quantitative impact of different variables on economic business cycles. Tinbergen combines Yule's approach of multiple regression (see chapter 6, section 3) with a dash of Fisher's views on inference. There is no reference to von Mises' *Kollektiv* or the Neyman–Pearson theory of inductive behaviour, and the epistemological interpretations of probability are totally absent. Although the title of his book, *Statistical Testing of Business Cycle Theories*, suggests otherwise, there is little testing of rival theories. Indeed, a statistical apparatus to test competing, non-nested theories simply did not exist. The few reported statistical significance tests are used to test the 'accuracy of results' (Tinbergen, 1939b, p. 12). They are not used as tests of the theories under consideration.

Tinbergen is interested in importance testing, in his words: testing the economic significance. Namely, he investigates whether particular effects have a plausible sign and are quantitatively important. If so, significance tests are used to assess the statistical accuracy or weight of evidence.[12] The procedure for 'testing' theories consists of two stages. First, theoretical explanations for the determination of single variables are tested in single regression equations. Second, the equations are combined in a system and it is considered whether the system (more precisely, the final reduced-form equation obtained after successive substitution) exhibits the fluctuations that are found in reality. In order to appraise theories, Tinbergen combines the most important features of the different theories in one 'grand model' (in this sense, as Morgan, 1990, p. 120, notes, Tinbergen views the theories as complementary rather than rivals: it is a precursor to comprehensive testing). Tinbergen's inferential results are, for example, that interest does not really matter in the determination of investment or that the acceleration mechanism is relatively weak. These are measurements, not tests.

The final goal of Tinbergen's exercise is to evaluate economic policy. This is done by using the econometric model as an experimental economy (or 'computational experiment', in modern terms). The coefficients in the model may change as a result of changing habits or technical progress, but Tinbergen (1939b, p. 18) writes,

> It should be added that the coefficients may also be changed as a consequence of policy, and the problem of finding the best stabilising policy would consist in finding such values for the coefficients as would damp down the movements as much as possible.

Keynes argues that Tinbergen does not provide the logical foundations which would justify his exercise. Keynes' critique is directly related to the *Treatise on Probability* and its emphasis on the principle of limited independent variety as an essential premise for probabilistic inference. There must be a certain amount of 'homogeneity' in order to validate induction. A major complaint of Keynes ([1939] CW XIV) is that Tinbergen's (1939a) particular example, the explanation of investment, is the least likely one to satisfy this premise, in particular, because of the numerous reasons that change investors' expectations about the future, a crucial determinant of investment. The method of mathematical expectations is bound to fail (Keynes, [1939] CW XIV, p. 152). Induction is justified in certain circumstances, only 'because there has been so much repetition and uniformity in our experience that we place great confidence in it' (Keynes, [1921] CW VIII, p. 290). Those circumstances do not apply generally, and Keynes' attitude towards probabilistic inference in the context of economics is very cautious.

Another complaint of Keynes deals with Tinbergen's choice of time lags and, more generally, the choice of regressors. In the rejoinder to Tinbergen (1940), Keynes ([1940] CW XIV, p. 319) suggests a famous experiment:

> It will be remembered that the seventy translators of the Septuagint were shut up in seventy separate rooms with the Hebrew text and brought out with them, when they emerged, seventy identical translations. Would the same miracle be vouch-

safed if seventy multiple correlators were shut up with the same statistical material?

This rhetorical question has been answered by the postwar practice of empirical econometrics: there is no doubt that specification uncertainty and the coexistence of rival, conflicting theories are deep problems that have not yet been solved. Friedman (1940, p. 659), in a review of Tinbergen's work, concurs with Keynes' objection:

> Tinbergen's results cannot be judged by ordinary tests of statistical significance. The reason is that the variables with which he winds up... have been selected after an extensive process of trial and error *because* they yield high coefficients of correlation.

Friedman adds a statement that reappears many times in his later writings: the only real test is to confront the estimated model with new data.

David Hendry (1980, p. 396) is not convinced that Keynes' objections are fundamental ones:

> Taken literally, Keynes comes close to asserting that no economic theory is ever testable, in which case, of course, economics ceases to be scientific – I doubt if Keynes intended this implication.

In fact, Keynes did. As Lawson (1985) also observes, Keynes is very clear on the difference between the (natural) sciences and the moral 'sciences', such as economics, and indeed believes that (probabilistic) testing is of little use in as changing an environment as economics. Keynes ([1921] CW VIII, p. 468) concludes his *Treatise* with some cautious support for applying methods of statistical inference to the *natural* sciences, where the prerequisites of fruitful application of statistical methods are more likely to be satisfied:

> Here, though I have complained sometimes at their want of logic, I am in fundamental sympathy with the deep underlying conceptions of the statistical theory of the day. If the contemporary doctrines of biology and physics remain tenable, we may have a remarkable, if undeserved, justification of some of the methods of the traditional calculus of probabilities.

Applying frequentist probability methods in economics not only violates logic, but is also problematic because of the absence of 'limited independent variety' in what would constitute the potentially most interesting applications, such as business cycle research. In her discussion of the debate, Morgan (1990, p. 124) claims that '[b]ecause Keynes "knew" his theoretical model to be correct, logically, econometric methods could not conceivably prove such theories to be incorrect'. Keynes may be accused of arrogance, but it is wrong to argue that he is as dogmatic as suggested here. Morgan's admiration for Tinbergen blinds her to the analytical argument of Keynes. The assertion that Tinbergen and other 'thoughtful econometricians' (p. 122) were already concerned with his logical point does not hold ground: the debate between Tinbergen and Keynes shows that they talk at cross purposes. Tinbergen never seriously considers the analytical problem of induction or the justification of sampling methods to economics. In Tinbergen's defence, one might note that he was primarily interested in economic importance tests and much less in probabilistic significance tests. Hence, a probabilistic foundation was not one of his most pressing problems.

3 Carnap and confirmation theory

3.1 Confirmation

Rudolf Carnap has tried to supersede Keynes' theory of logical probability by proposing a quantitative system of inductive probability logic, without endorsing Keynes' incommensurability thesis or his principle of limited independent variety (see Carnap, 1963, pp. 972–5). Carnap was one of the most important members of the *Wiener Kreis*, and he made many fundamental contributions to the philosophy of logical positivism. Initially, he and his colleagues of the *Wiener Kreis* supported the (von Mises branch of the) frequentist theory of probability. Keynes' theory was thought to lack rigour, it was 'formally unsatisfactory'. Around 1941, Carnap reconsidered his views. His interest in the logical theory of probability was raised, due to the influence of Wittgenstein and Waissman. He began to appreciate Keynes' and Jeffreys' writings (p. 71; see Schilpp, 1963, for an overview of Carnap's philosophy).

In his early writings, Carnap attempted to show that theoretical statements have to be based (by definition) on experience in order to be 'meaningful'. The verifiability principle (see chapter 1, section 5.1) can then be used in order to demarcate 'science' from 'metaphysics'. Popper pointed to the weaknesses of the verifiability principle. Instead of giving up his positivist theory of knowledge, Carnap amended his theory by relaxing the verifiability principle. Here, I will deal with Carnap's amended theory of knowledge.

Definitions based on experience do not yield meaningful knowledge, but the possibility of (partial) confirmation does. The basis for Carnap's theory (Carnap, 1950; 1952) is an artificial language of sentences along the lines of Russell and Whitehead's *Principia Mathematica*. The goal is to construct a confirmation function that can be used for inductive inference, which can figure as a common base for the various methods of

Fisher, Neyman–Pearson and Wald for estimating parameters and testing hypotheses (Carnap, 1950, p. 518).

A confirmation function $c(H, A)$ denotes the degree of confirmation of an hypothesis H by the evidence A. Carnap interprets this as the quantitative concept of a logical probability. He shows that a conditional probability $P(H|A)$ (in the simplest case defined on a monadic predicate language $L(A, n)$ which describes n individuals a_n $(i = 1, \ldots, n)$ who have or have not characteristic A) is indeed a measure of this degree.

The problem is to find the necessary prior probability by means of logical rules. For the given simple case, there are 2^n possible states or sentences. One solution is to invoke the principle of indifference, but this does not yield unique prior probabilities. Carnap (1950) proposes two logical constructs, $c^\dagger$(which bases the prior probability in the example on 2^{-n}) and c^* (where, for $0 < q < n$, the prior probability depends on the number of combinations ${}^nC_q$). These constructs are discussed in Howson and Urbach (1989, pp. 48–55). Both rely on considerations of symmetry. If $n \to \infty$, the logical probability shrinks to zero. In more complex examples it will be difficult to construct similar prior probabilities. If the system is extended to a general 'universe of discourse' (or sample space), then it is unclear how the logical probabilities should be constructed. Without further empirical information, the logically possible states of the world have to be mapped into prior probability assignments, but how this should be done remains unclear.

In a critical book-review of Carnap (1950) in *Econometrica*, Savage (1952, p. 690) concludes that until the mapping problem is resolved, 'application of the theory is not merely inadequately justified; it is not even clear what would constitute an application'. Not surprisingly, Carnap's work did not gain a foothold in econometrics.[13] The unifying base for the various methods of estimating and testing was not recognized by the econometric profession. Another critique has been expressed by Popper ([1935] 1968, p. 150), who complains that the resulting probability statements are non-empirical, or tautological. This is true for the construction of priors, but it is not correct that Carnap has no theory of learning from experience.

3.2 The generalized rule of succession

Carnap's theory is a generalization of Laplace's rule of succession, discussed in chapter 2, section 3. Let H and A be 'well formed formulas' describing an hypothesis H and evidence A for a given predicate language L. In the case of casting a die, A^k denotes the outcomes of k throws of a die. The hypothesis H_i states that the next throw will be an outcome A_i.

In the evidence, k' throws have characteristic A_i. Carnap's degree of confirmation is summarized in the following equation (a derivation can be found in Kemeny, 1963, pp. 724–9):

$$c(H_i, A^k) = \frac{k' + \frac{\lambda}{\kappa}}{k + \lambda}, \qquad (3)$$

where κ is the number of exclusive and exhaustive alternatives (e.g. $\kappa = 2$ in a game of heads and tails; $\kappa = 6$ in case of throwing dice). The fundamental problem is how to choose λ, an arbitrary constant which obeys $0 \leq \lambda \leq \infty$. It represents the logical factor by which beliefs are determined. The problem of choosing λ has not been solved, as a unique optimal value for λ does not exist. If the game is heads and tails and λ is set equal to 2, then Laplace's rule of succession obtains. If λ has a higher value, more evidence is needed to change one's opinion; if λ grows to infinity, empirical evidence will not influence probability assignments. On the other hand, setting $\lambda = 0$ yields the 'straight rule'. In other words, λ measures the unwillingness to learn from experience. It quantifies the trade-off between eagerness to increase knowledge, and aversion to taking risk by attaching too much weight to the first n outcomes of an experiment.

The problem also shows up in Bayesian econometrics, where a choice has to be made about the relative weights of prior and posterior information before posterior probabilities, predictive densities, etc. can be calculated. In a discussion of some possible objections to using uniform prior probabilities, Berger ([1980] 1985, p. 111) gives an example of an individual trying to decide whether a parameter θ lies in the interval (0,1) or [1,5). The statement of the problem suggests that θ is thought to be somewhere around 1:

> but a person not well versed in Bayesian analysis might choose to use the prior $\pi(\theta) = 1/5$ (on $\Theta = (0, 5)$), reasoning that, to be fair, all θ should be given equal weight. The resulting Bayes decision might well be that $\theta \in [1, 5)$, even if the data moderately supports (0,1), because the prior gives [1,5) four times the probability of the conclusion that $\theta \in (0, 1)$. The potential for misuse by careless or unscrupulous people is obvious.

The warning is odd. There is no *logical* reason why such a choice is careless or unscrupulous. If there is good reason for the choice of the prior, it should have some weight in the posterior. The only thing a Bayesian can do is to make a trade-off between the thirst for knowledge and the aversion to risk (which happens to be one of the major goals of Berger's book).

Hintikka has extended Carnap's theory to more elaborate languages (polyadic, infinite). He has solved some analytical problems. Hintikka is able to reject Carnap's conclusion that universal statements that all individuals of a countable universe have some specific property must have a prior and posterior probability equal to zero. Carnap's conclusion has been more strongly expressed by Popper, who has tried to prove that the logical probability of scientific laws must be zero. The fundamental problem, how to settle for appropriate prior probabilities on logical grounds, has not been solved by Hintikka (see Howson and Urbach, 1989, and the references cited there).

3.3 An assessment

Efforts to establish prior probabilities on purely philosophical grounds have been of limited success. More recently, however, one of Carnap's students, Ray Solomonoff, introduced information-theoretical arguments to analyse the problem. In combination with earlier intuitions of Harold Jeffreys, this led to a breakthrough in the literature on prior probabilities and inductive inference. Section 5 of this chapter discusses various more recent contributions to the theory of probability *a priori*.

The practical compromise of inference must be to select priors that are robust in the sense that the posterior distribution is relatively insensitive to changes in the prior. *How* insensitive is a matter of taste, like the choice of λ is. The personalist theories of inference try to provide a deeper analysis of the choice between risk aversion and information processing.

Before turning to that theory, a final issue of interest in Carnap's work deserves a few words. This is the distinction between confirmability and testability. A sentence which is confirmable by possible observable events is testable if a method can be specified for producing such events at will. Such a method is called a test procedure for the sentence. This notion of testability corresponds, according to Carnap, to Percy Bridgman's principle of operationalism (which means that all physical entities, processes and properties are to be defined in terms of the set of operations and experiments by which they are apprehended). Carnap prefers the weaker notion of confirmability to testability, because the latter demands too much. It seems that in econometrics, confirmability is the most we can attain as well (see Carnap, 1963, p. 59). This will be illustrated in chapter 8 below.

4 Personalist probability theory

The formal foundations of the personalist theory of probability are due to the French radical socialist and probability theorist Emile Borel (his paper of 1924 can be found in Kyburg and Smokler, [1964] 1980) and Keynes' pupil Frank Ramsey (1926). They base the theory of probability on human ignorance, not infinite sequences of events. Both hold that probability should be linked with rational betting behaviour and evaluated by studying overt behaviour of individuals. They differ from Reichenbach (1935; [1938] 1976) in making subjective betting ratios the primitive notions of their theories of probability, whereas Reichenbach starts from relative frequencies and introduces the wager as an afterthought (see, e.g., pp. 348–57, for his discussion of Humean scepticism, induction and probability). Like Neyman and Pearson, the early personalist probability theorists were influenced by the upcoming theory of behaviourism.

Borel's legacy to econometrics is primarily the Borel algebra. His other contributions to probability theory and mathematics are little known in the econometric literature. Ramsey is still often cited in economic theory, but hardly ever in econometrics. The philosophical writings of Borel and Ramsey did not have an appreciable influence on econometrics. That is not entirely true for two more recent contributors to the personalist theory of probability: de Finetti and Savage. Their work is discussed in the following sections.

4.1 De Finetti, bookmaker of science

The key to every constructive activity of the human mind is, according to Bruno de Finetti (1975, p. 199), Bayes' theorem. De Finetti belongs to the most radical supporters of the personalist theory of probability. De Finetti ([1931] 1989, p. 195) defends a strictly personalistic interpretation of probability: 'That a fact is or is not practically certain is an opinion, not a fact; that I judge it practically certain is a fact, not an opinion.' He outlines a theory of scientific inference, which is further developed in more recent work (de Finetti, 1974; 1975). He proposes to escape the straitjacket of deductive logic, and instead holds subjective probability the proper instrument for inference. Furthermore, he rejects scientific realism. In the Preface to his *Theory of Probability*, de Finetti claims (in capitals) that PROBABILITY DOES NOT EXIST. Probability does neither relate to truth, nor does it imply that probable events are indeterminate (de Finetti, [1931] 1989, p. 178; 1974, p. 28). Probability follows from uncertain knowledge, and is intrinsically subjective. It should be formulated in terms of betting ratios.

There is one 'objective' feature that probability assessment should satisfy: coherence, which is a property of probability assignments:

Coherence. A set of probability assignments is coherent, if and only if no Dutch Book (a bet that will result in a loss in every possible state of the world) can be made against someone who makes use of these probability assignments.

Coherence is de Finetti's requirement for rationality. The probability assignments of a rational individual (someone who behaves coherently) are beyond discussion, as de Finetti's (1974, p. 175) theory is strictly personalistic: 'If someone draws a house in perfect accordance with the laws of perspective, but choosing the most unnatural point of view, can I say that he is wrong?' His answer is 'no'.

The Dutch Book argument is one of the pillars of modern Bayesianism. It is a natural requirement to economists, many economic theorists accept Bayesian probability theory for this reason. Econometricians are less convinced: its applicability to empirical research is limited (see also Leamer, 1978, p. 40). Typically, it will be difficult to make a bet about scientific propositions operational. Worse, it may be undesirable or without scientific consequences. The more operational the bet, the more specific the models at stake will be, and hence, the less harmful a loss will be for the belief of the loser in his general theory. Models are not instances of theories, at least in economics. A related critique of book-makers' probability has been expressed by Hilary Putnam (1963, p. 783):

> I am inclined to reject all of these approaches. Instead of considering science as a monstrous plan for 'making book', depending on what one experiences, I suggest that we should take the view that science is a method or possibly a collection of methods for selecting a hypothesis, assuming languages to be given and hypotheses to be proposed.

Good scientists are not book-makers, they are book writers. The question remains how fruitful hypotheses emerge, and how they can be appraised, selected or tested. Putnam does not answer that question. Neither does de Finetti.

De Finetti follows Percy Bridgman in requiring an operational definition of probability (1974, p. 76). By this he means that the probability definition must be based on a criterion that allows measurement. The implementation is to test, by means of studying observable decisions of a subject, the (unobservable) opinion or probability assignments of that subject. Hence, to establish the degrees of beliefs in different theories of scientists, their observable decisions in this respect must be studied. Whether this can be done in practice is questionable.

4.2 Exchangeability

An interesting aspect of de Finetti's work is his alternative for the formal reference scheme of probability, discussed in the Intermezzo. One of the basic notions of probability is independence, defined as a characteristic of events. For example, in tossing a fair coin, the outcome of the nth trial, x_n (0 or 1), is independent of the outcome of the previous trial:

$$P(x_n|x_{n-1}, x_{n-2}, \ldots) = P(x_n). \tag{4}$$

This is a purely mathematical definition. De Finetti wants to abandon this characteristic, because in case of independence, the positivist problem of inference (learning from experience) gets lost: if A and B are independent, then $P(A|B) = P(A)$. Therefore, he introduces an alternative notion, exchangeability of events. It is defined as follows (de Finetti, 1975, p. 215):

Exchangeability. For arbitrary (but finite) n, the distribution function $F(\cdot, \cdot, \ldots, \cdot)$ of $x_{h1}, x_{h2}, \ldots, x_{hn}$, is the same no matter in which order the x_{hi} are chosen. Hence, it is not asserted that $P(x_i|x_{h1}, x_{h2}, \ldots, x_{hn}) = P(x_i)$, but

$$\begin{aligned} &P(x_i|x_{h_1}, \ldots, x_{h_n}) = P(x_j|x_{k_1}, \ldots, x_{k_n}) \\ &\qquad \forall(i, h_1, \ldots, h_n) \wedge (j, k_1, \ldots, k_n), \end{aligned} \tag{5}$$

as long as the same number of zeros and ones occur in both conditioning sets.

This is in fact the personalist version of von Mises' condition of randomness, with the difference that de Finetti has finite sequences in mind. Using this definition, de Finetti is able to establish the convergence of opinions among coherent agents with different priors. Note that it is not yet clear how to ascertain exchangeability in practice. Like von Mises' condition of randomness in the frequency theory of probability, the exchangeability of events must finally be established on subjective grounds.

4.3 The representation theorem

A by-product of exchangeability is the representation theorem. It is the proposition that a set of distributions for exchangeable random quantities can be represented by a probabilistic mixture of distributions for the same random quantities construed (in the objectivist, relative frequency way) as independent and identically distributed. If the probability $P(x_1, \ldots, x_n)$ is exchangeable for all n, then there exists a prior probability distribution $\pi(\lambda)$ such that

$$P(x_1, \ldots, x_n) = \int_0^1 \lambda^k (1-\lambda)^{n-k} \pi(\lambda) d\lambda. \tag{6}$$

This theorem provides a bridge between personalistic and objective interpretations of probability. Exchangeability enables de Finetti to derive Bernoulli's law of large numbers from a personalistic perspective. Furthermore, if $\pi(\lambda) = 1$, it is possible to derive Laplace's rule of succession from the definition of conditional probability and the representation theorem.

This is an important theorem, but there are some unsolved questions. A technical problem is that a general representation theorem for more complicated cases (such as continuous distributions) is not available. A more fundamental problem is how can two personalists, who have different opinions about a probability, converge in opinion if they do not agree about the exchangeability of events in a given sequence? There is no 'objective' answer to this question.

4.4 Savage's worlds

Savage's theory is an extension of de Finetti's work (Savage, [1954] 1972, p. 4). They co-authored a number of articles and share the personalistic perspective, but some differences remain. Savage bases his theory on decision making (like de Finetti) but combines probability theory more explicitly with expected (von Neumann–Morgenstern) utility theory.[14] De Finetti does not directly rely on utility theory, which makes his view vulnerable to paradoxes such as Bernoulli's 'St. Petersburg paradox', analysed in Savage (p. 93). Personalistic probability and utility theory are the two key elements of Savage's decision theory. Savage continues the tradition of Christiaan Huygens and Blaisse Pascal, who were the first to derive theorems of expected utility maximization. Another particular characteristic of Savage is his support of minimax theory (which is due to von Neumann, introduced in statistics by Abraham Wald). Savage (p. iv) notes that 'personalistic statistics appears as a natural late development of the Neyman–Pearson ideas'.

Savage's decision theory is positivistic in the sense that it is about observable behaviour (but Savage acknowledges that real human beings may violate the perfect rationality implied by his theory). He does not analyse preference orderings of goods, or events, but orderings of acts. You may assert preference of a Bentley over a Rolls Royce, but this is meaningless if you never have the opportunity to make this choice in real life.

Of particular interest among Savage's ideas about inference, is his notion of small and large worlds (pp. 82–91). It is perhaps the most complicated part of his book, and he acknowledged that it is 'for want of a better one' (p. 83) (readers who want to avoid a headache might skip the remainder of this subsection). The intuition of the idea is simple: the analytical straitjacket can be imposed on specific or 'local' problems (the small world). However, when it turns out that the straitjacket really does not fit the problem, one has to shop in a larger world and buy a new straitjacket which is better suited to the issue at stake.

Savage begins by defining a world as the object about which a person is concerned. This world is exhaustively described by its states $s, s', \ldots$. Events $A, B, C, \ldots$ are defined as sets of states of the world. They are subsets of S, the universal event, or the *grand world*. It is a description of all possible contingencies. The true state of the world is the state that does in fact obtain.

Consequences of acting are represented by a set F of elements $f, g, h, \ldots$. Acts are arbitrary functions f, g, h from S to F. Act g is at least as good as or preferred to act f (i.e. $f \preceq g$) if and only if:

$$\mathcal{E}(f - g) \leq 0, \qquad (7)$$

where E denotes the expected value of an act, derived from a probability measure P. Given the axioms of choice under uncertainty, an individual has to make once in his lifetime one grand decision for the best act, contingent on all possible states of the world. Similarly, ideally an econometrician has to know exactly how to respond to all possible estimates he makes in his effort to model economic relationships, and an ideal chess player would play a game without further contemplation once the first move is made. In practice, this is infeasible. Problems must be broken up into isolated decision situations. They are represented by *small worlds*.

A small world is denoted by $\bar{S}$. It is constructed from the grand world S by partitioning S into subsets or small world states, $\bar{s}, \bar{s}', \ldots$. A small world state is a grand world set of states, i.e. a grand world event. In making a choice between buying a car and a bicycle, in a small world one may contemplate contingencies such as the oil price and the expected increase in traffic congestion. The large world contingencies are many, perhaps innumerably many. Take, for example, the probability of becoming disabled, the invention of new means of transportation, but also seemingly irrelevant factors such as changes in the price of wheat or the average temperature on Mars. The idea of the construction of a small world is to integrate all 'nuisance' variables out of the large world model.

A small world event is denoted by $\bar{B}$, which is a set of small world states in $\bar{S}$. The union of elements in $\bar{B}$, $\cup\bar{B}$, is an event in S. A small world is 'completely satisfactory' if and only if it satisfies Savage's postulates of preference orderings under uncertainty, and agrees with a probability $\bar{P}$ such that

$$\bar{P}(\bar{B}) = P(\cup\bar{B})\ \forall\bar{B} \subset \bar{S}, \tag{8}$$

and has a small world utility $\bar{U}(\bar{f})$ of consequences of acts $\bar{f}$ which is equal to the large world expected value of the small world consequences of acts, $E(\bar{f})$. This small world is called a *microcosm*. The probability measure $\bar{P}$ is valid for the small world. In words, the grand-world probability P of the union of elements in the small-world event $\bar{B}$ is equal to its small-world probability $\bar{P}$. If a small world gives rise to a probability measure and utility consistent with those of the grand world, the small world is said to be completely satisfactory. The problem, of course, is to find out when the conditions are fulfilled. On this question, Savage has little to say.

4.5 *Inference, decision and Savage's impact on econometrics*

Savage's theory of rational conduct is in conflict with reality. Tversky and Kahneman (1986) argue that no theory of decision can be normatively adequate and descriptively accurate. Savage is aware of conflicts between theory and reality, and opts for a normative theory. The theory is still frequently used to analyse human conduct, but as a statistical theory of inference it is less successful. It is doubtful that 'acts' of econometricians (or scientists in general), for example the choice of models, have consequences which can be measured in terms of utility (and are worth the measurement), even in a small world context. This may explain why Savage's impact on econometrics has been small. Although he was in Chicago (from 1950–1954) when the Cowles Commission resided there, he did not leave a mark on the direction of econometric research at Cowles.[15]

5 The construction of prior probabilities

A priori probabilities are indispensable to inductive inference, but their justification is hard. Where do they come from? From previous experience? Are they logical constructs? Or are they just subjective whimsical creations of our imagination? If the specification of prior probability results from previous applications of Bayes' theorem, then the question carries over to the justification of such previous inferences, leading to infinite regress. 'In the beginning', there was ignorance. However, priors representing ignorance are not always easy to find.

The third option, that priors are subjectively determined, seems to be the most attractive one. There are some restrictions to the choice of a prior:

- coherence,
- communicability, and
- tractability.

Coherence is de Finetti's requirement for rational judgement. The other two restrictions may be useful for individual behaviour as well, but are particularly important for scientific inference, which is based on inter-subjective scrutiny and the possibility of replication.

In many cases, explicit prior information is not available. Still, from a logical point of view, a prior representing ignorance must be available for a theory of inference (although there are occasionally Bayesians who renounce them, e.g. Leamer, 1978, pp. 61–3). Non-informative priors are required for the first step in inference. The following sections discuss alternative ways to construct priors, including non-informative priors.

5.1 Informative priors

If prior information is available and if this can be represented in a prior probability distribution, then there are in principle no conceptual difficulties for applying Bayes' theorem. This is acknowledged by all statisticians alike, not only Bayesians. Some popular prior probability distributions are of the natural conjugate form, in which case the posterior distribution is of the same family of distributions as the prior distribution. Analytical tractability is an important advantage of such priors, at the cost of flexibility (Leamer, 1978, p. 79). The revolution in computation techniques has generated interest in more complex prior probability distributions. Numerical integration techniques can be applied if analytical derivation of posterior probabilities is impractical or impossible.

The information imputed in informative prior probability distributions may result from theoretical considerations (e.g. if a parameter is constrained to a certain interval, like budget shares in consumer demand), or from earlier experience. An alternative approach is to estimate the prior probability distribution from the raw data. The 'empirical Bayes' method uses this approach, but applications have been relatively scarce and are primarily found in problems of statistical quality control. From a philosophical point of view, empirical Bayes methods add little to an understanding of inductive inference.[16]

5.2 Indifference again: non-informative priors and the uniform distribution

The oldest method to construct a prior density, due to Bayes and Laplace, is to construct a *uniform* prior probability distribution.[17] The indifference theory of probability, discussed in chapter 2, introduced this approach. For example, Bayes used a uniform density to solve his original problem. A problem is that this choice may lead to paradoxical results (e.g. Bertrand's paradox – chapter 2, section 2.2), although the paradoxes may be overcome by including all relevant dimensions of indifference. Another objection might be that it violates the fourth axiom of probability, given in the Intermezzo, namely that the total probability should be equal to one, not exceed it. This is why this prior is called improper. The posterior may still satisfy the axioms; using improper priors is, therefore, not prohibited. Jeffreys ([1939] 1961, p. 21) holds that the fourth axiom is just a convention which is often useful but sometimes not.

Harold Jeffreys brought the method of inverse probability back to the forefront of statistics, among other things for his significant contributions to the construction of non-informative prior probabilities. Whereas Keynes was primarily concerned with the logic of inference, Jeffreys was more interested in statistical applications. This was at a time when, among statisticians, the methods of Bayes and Laplace were almost universally rejected. The problem of deriving prior probability distributions was thought to be insurmountable. But Jeffreys was able to provide a better justification for the use of non-informative priors than Bayes and Laplace had done. Like Carnap, Jeffreys was inspired by the *Principia Mathematica* of Russell and Whitehead: Jeffreys tried to make probability theory axiomatic as a theory of scientific inference. Unlike

Ramsey and Savage, Jeffreys does not combine utility theory and probability theory (see e.g. Jeffreys, [1939] 1961, pp. 30–3).

The problem of the uniform prior probabilities used by Bayes and Laplace is that they are not invariant to non-linear transformations, and that they are not appropriate for representing ignorance if the parameter for which the prior probability is construed is one-sided unbounded. But consider first the *two*-sided unbounded location parameter (or mean) λ and data x. Like Bayes and Laplace, Jeffreys (p. 119) recommends a uniform prior,

$$P(\lambda) = \text{constant}. \tag{9}$$

The justification for using this prior distribution is that it tends to give 'satisfactory' results (although it undermines significance testing, therefore, Jeffreys later introduces the 'simplicity postulate', discussed in chapter 5 below). Furthermore, it has the advantage that the probability ratio of two segments of the uniform distribution is indeterminate, which appeals to 'being ignorant'.

The *one*-sided bounded parameter is where Jeffreys made his own contribution. It is easy to see that a uniform distribution for such a parameter is not justified. Take the scale parameter σ, $0 \leq \sigma < \infty$. Assign a uniform probability to σ. Then for some finite value of σ, say σ^*, the probability that $\sigma < \sigma^*$ will be finite, whereas the probability that $\sigma > \sigma^*$ is infinite. This is not a satisfactory representation of prior ignorance. Therefore, an alternative to the uniform distribution over σ is a uniform distribution over $\partial\sigma/\sigma$, which results in

$$P(\sigma) \propto (1/\sigma), \tag{10}$$

where $\propto$ denotes 'proportional to'. An advantage is that this expression is proportional to non-linear transformations of σ (for example, $\partial \ln \sigma \propto \partial \ln \partial^2$ which implies that ignorance on the scale parameter means ignorance on the variance). A generalization of this notion of 'invariance' is given in Jeffreys (pp. 179–92), which has generated a large literature on invariance properties of probability distributions.

Jeffreys' rules are based on pragmatic insights. Critics argue that, therefore, his methods lack justification, like the earlier work of Bayes and Laplace. A second objection to Jeffreys' prior is that it is not independent from experience: the sampling model is used for its construction (and, as is frequently the case in econometrics, the sampling model may be the result of a 'fishing' or data-mining expedition). The maximum entropy prior, discussed below, avoids this unjustified dependency on the data and can be derived from more general principles.

A third problem is that improper priors may lead to a 'marginalization' paradox (Dawid, Stone and Zidek, 1973). Consider two Bayesian statisticians, who are given the same data $x = (y, z)$. The problem involves the inference on parameters $\theta = (\eta, \xi)$ where ξ is the parameter of interest. The first statistician uses an improper prior on θ, and marginalizes subsequently. The second investigator, who starts directly from the simple problem (without regard of y and η), uses a proper prior for ξ. The paradox is the fact that the second investigator is unable to reproduce the inference of the first one, whatever prior on ξ is used. The conclusion of this work is usually interpreted as a warning against the use of improper priors. However, Jaynes (1980) shows that the paradox does not result from the use of improper priors, but from an unjustified neglect of differences in information sets of the two statisticians.

Jeffreys contributions did not make an immediate impact on the development of econometrics, although it may have been a 'near miss'.[18] Only much later, Zellner popularized Jeffreys' Bayesian approach in econometrics.

5.3 The maximum entropy principle

Information theory provides a way to deal with uncertainty, which is called disorder or entropy. Information theory is related to probability theory and indeed has progenitors in probability theory (Laplace) and in statistical mechanics (Ludwig Boltzmann, Josiah W. Gibbs). The concept was mentioned in a discussion of utility theory in Davis' (1941) *Theory of Econometrics*. But the explicit development is due to the information theory of Claude Shannon and the probabilistic elaboration of Jaynes (1957). Recently, entropy has received more interest from econometricians who have used it for the problem of model choice and testing rival theories (Maasoumi, 1987, discusses econometric implementations of entropy).

Shannon's basic idea is the notion that the occurrence of events reveals information. If an event with high probability materializes, then little information is gained. The *information function* measures the information gain by comparing the prior probability with the posterior probability of the event. Popper uses this to measure his degree of corroboration (see chapter 3), and measures of divergence (or distance) follow directly from the notion of expected information.

Define $h(P)$ as the information revealed by the occurrence of an event with probability P, $h(P)$ being a decreasing function of P. It turns out that $h(P) = \log(1/P)$ is a convenient measure of information. The expected value of the information content is the entropy. For the discrete case, with event outcomes A_i ($i = 1, \ldots, n$), the entropy of the according probability density $P(A)$, denoted by $\text{Ent}(P(A))$, is given by:

$$\text{Ent}(P(A)) = \sum_{i=1}^{n} P(A_i) \log \frac{1}{P(A_i)}, \tag{11}$$

and is zero by definition if $P(A_i)$ is zero for all i. This is a unique representation of a consistent information measure in case of discrete probability distributions. The maximum entropy principle (MEP; due to Jaynes, 1957) is to maximize this expression subject to the restriction

$$\sum_{i=1}^{n} P(A_i) = 1. \tag{12}$$

If other information is available (empirical evidence, auxiliary constraints), then it should be used for additional constraints in the maximization problem sketched above (each constraint is a piece of information). MEP has a clear relation to the efficient market theorem that is familiar in economics. MEP states that all available information should be used for problems of inference (the efficient market theorem goes one step further: all available information *will* be used).

The entropy has its maximum value if all probabilities for different A_i are equal, i.e. $P(A_i) = 1/n$ in which case $\text{Ent}(P(A)) = \log n$. For dice, coins and other symmetric games of chance, the principle of maximum entropy leads to the same prior probability distributions as obtained from the principle of indifference or Jeffreys' prior. However, Jaynes uses MEP to provide a better general justification for the use of prior probability distributions representing ignorance. MEP is of further use to find out if there are additional constraints that should be considered, inspired by a divergence between the maximized entropy without such further constraints, and the observed entropy measure.

The modelling strategy implied by MEP is from simple to general. It does not depend on a 'true' probability distribution, the necessary ingredient of frequency theories of probability. Jaynes (1978, p. 273) objects to the frequency interpretation: 'The Principle of Maximum Entropy, like Ockham, tells us to refrain from inventing Urn Models when we have no Urn.' A final argument in favour of the MEP is the statement that, in two problems with the same prior information, the same prior probabilities should be assigned (p. 284). This makes it a 'necessarist' approach, contrary to the subjective one of de Finetti and Ramsey.

The maximum entropy principle is a natural way to formulate hypotheses for problems with symmetry properties. In a problem of throwing a die, without prior information on the quality of the die, the maximum entropy principle leads to assigning a prior probability of $P(A_i) = 1/6$ to all six faces of the die. If a large series of castings of a die generate relative frequencies that are approximately equal to 1/6 as well, leading to a posterior probability $Q(A_i) = 1/6$, then hardly any information is gained. The information gain is defined as:

$$\text{Ent}(Q(A_i)) - \text{Ent}(P(A_i)) = \log \frac{Q(A_i)}{P(A_i)}. \tag{13}$$

The *expected* (average) information gain is:

$$I(Q, P) = \sum_{i=1}^{n} Q(A_i) \log \frac{Q(A_i)}{P(A_i)}. \tag{14}$$

This definition of expected information generalizes to the concept of divergence, which is useful for measuring the difference in information contained in two rival descriptions (distribution functions) of the data.

5.4 Universal priors

A recent development in computer science and mathematics is the application of Kolmogorov complexity theory to inductive inference and machine learning. This results, among other things, in a generalization of the maximum entropy priors. Furthermore, it gives new insights into the use of simplicity in inference (and on the relation between the simplicity of hypotheses and their prior probabilities). The method of inductive reasoning, implicit in Rissanen (1983) and surveyed in Li and Vitnyi (1990; 1992), follows the idea of Solomonoff (1964a, b) to construct a universal prior which can be substituted for any particular 'valid' prior probability distribution in the set of computable probabilities. This can be used in applications of Bayes' theorem.

Solomonoff (1964a) shows that, in theory, a universal prior exists and generates approximately as good results as if the actually valid distribution were used. A problem is that the universal prior itself is not computable as a result of the halting problem (the possibility that a computer is unable to 'decide' whether a problem can be solved; see chapter 5, n. 8).

This would impose a serious limitation on an idealized theory of computerized inference, based on a mechanized learning scheme that exploits Bayes' theorem to its limits. Recent advances in complexity theory suggest however computable approximations to this universal prior. The theory of universal priors purports to derive 'objective' (or logical) prior probability distributions by means of algorithmic information theory. Because the notion of universal priors is strongly related to another topic of interest, i.e. simplicity, the discussion will be continued in the following chapter on the theory of simplicity.

6 Appraisal

This chapter has discussed epistemological approaches to inference, based on Bayes' theorem. Bayesian probability theory encompasses views that vary in their analysis of limitations of the human mind. Necessarist theories, following Keynes, Carnap, and to a lesser degree Jeffreys, are closest to an ideal type of perfect inference. But even the necessarists discuss cognitive limitations to human inference. In particular, Keynes argues that often logic will be insufficient for eliciting numerical probabilities to propositions. Instead of concluding that, therefore, human behaviour has to be erratic and society is like a madhouse, Keynes concludes that human beings do and should rely on convention and custom.

A different angle has been chosen by the personalists, like Ramsey, de Finetti and Savage. They do not worry about the problem of assessing logically correct prior probabilities. Instead, they regard priors as a matter of taste, and *de gustibus non est disputandum*. Probabilistic inference is personal, objective probability does not exist. This seems to introduce whims in inference, but those are constrained by a tight straitjacket: coherence. Again, not a madhouse, but once the prior is chosen there is no escape from de Finetti's prison. Savage has made an effort to create an emergency exit, by distinguishing small worlds (the personal prison) from the large world. He does not clarify when it is time to escape nor what direction to take from there.

Both logical and personal approaches to epistemological probability have been criticized for a lack of scientific objectivity. If scientific inference depends on personal taste or custom and convention, it loses much of its appeal. But objectivity is as remote from the frequency approach as it is from the Bayesian approach to inference. Taste (specification), convention (Fisher's 0.05), idealization (hypothetical population, infinitely large) and a separation of object and subject are the cornerstones of frequentist inference. Bayesianism is more explicit about judgements as input, whereas the frequentists sweep them under the carpet by calling assumptions knowledge, as Good (1973) argues. Objectivity is not the real battleground. The real dispute is about pragmatic methods and the robustness with respect to the choice of priors. Efron (1986, p. 2) complains that

> Bayesian theory requires a great deal of thought about the given situation to apply sensibly... All of this thinking is admirable in principle, but not necessarily in day-to-day practice. The same objection applies to some aspects of Neyman–Pearson–Wald theory, for instance, minimax estimation procedures, and with the same result: they are not used very much.

Simplicity of methods is important. Non-informative priors are useful ingredients in Bayesian inference, and in this case, inference based on maximum likelihood inferences quite often has a straightforward Bayesian interpretation. Apart from elaborating the Bayesian philosophy of science, Jeffreys has made an effort to make Bayesian methods more palatable, both as pragmatic methods, and as one that is not too sensitive on the choice of priors. He suggests circumventing Keynes' problem of absolute uncertainty by a system of non-informative priors that improves upon the less satisfactory priors of Bayes and Laplace. Jeffreys contributed the most important breakthrough in epistemological probability since Bayes and Laplace. The theory of universal priors is a small but important elaboration of Jeffreys' approach.

In practice, many econometricians interpret their work (which is obtained by using frequentist tools) in a Bayesian fashion. The temptation to attach (quasi-)probabilities to hypotheses is hard to resist. This Bayesian interpretation of frequentist results is, to some extent, justifiable by the asymptotic equivalence of many of frequentist and Bayesian inferences. In the long run, the sample information dominates the prior information (von Mises' 'second law of large numbers'). This Bayesian interpretation of frequentist econometric inferences has some drawbacks. First, particularly in macro-econometrics, samples are small and relatively uninformative. Then prior probability distributions do matter and their impact should be investigated and acknowledged. Second, inference on unit roots exemplifies a case where explicitly Bayesian methods seem to be preferable even in large samples (Sims and Uhlig, 1991). This is also a clear example where the choice of prior probability distributions is crucial. Finally, unlike frequentists, Bayesians claim that significance levels should vary with sample size. If this advice is not accepted, outcomes of frequentist and Bayesian tests will diverge.

Although the frequentist approach emphasizes its reliance on facts and nothing but the facts, the epistemological (Bayesian) approach to inference is more useful for a positivist methodology of economics. First, because of the logical fact that, in order to judge the support (or probability) for theories, Bayes' theorem is needed. Second, because of the fact that, in economics, there is no strong evidence for the validity of von Mises' convergence condition, to put it mildly. Measurement and prediction remain feasible, but only with less strong pretences. Epistemological inference serves this purpose.

Notes

1. When Keynes wrote the *Treatise*, 'events' was not yet a well-defined notion. Therefore, Keynes ([1921] CW VIII, p. 5) argues that it is more fruitful to analyse the truth and probability of propositions than of ambiguously defined events. Kolmogorov (1933) clarified this issue by defining events as subsets of the sample space.
2. Jeffreys and Keynes were both inspired by the Cambridge logician and probability theorist, William Ernest Johnson, who also taught Jeffreys' collaborator Dorothy Wrinch, and, I presume, Ramsey (see Jeffreys, [1939] 1961, p. viii; the obituary by Keynes in *The Times* of 15 January 1931; and the dedication in Good, 1965). In Keynes' ([1933] CW X, p. 349) words, Johnson 'was the first to exercise the epistemic side of logic, the links between logic and the psychology of thought'. Johnson was the closest friend of the economist Neville Keynes, father of Maynard. In 1907, Johnson was examiner of Keynes' dissertation on probability (together with Alfred North Whitehead; see Skidelski, 1983, pp. 69 and 182).
3. Jeffreys ([1939] 1961, p. viii) writes that, in his biographical essay on Ramsey, Keynes withdrew his assumption that probabilities are only partially ordered. In fact, it is not the partial ordering which Keynes dropped, but (and only to a limited extent) the concept of logical or necessary probability.
4. The question of eggs illustrates Keynes' (1983 CW XI, pp. 186–216) argument with Karl Pearson during 1910–11. Here, the issue was the influence of parental alcoholism on the offspring. Keynes (p. 205) claims that the 'methods of "the trained anthropometrical statistician" need applying with much more care and caution than is here exhibited before they are suited to the complex phenomena with which economists have to deal'.
5. Keynes' definition of homogeneity (insofar as he really defines it) is different from the contemporary statistical definition (e.g. Mood, Graybill and Boes, 1974), which refers to the equality of probability distributions. Their existence is a presumed part of the 'maintained hypothesis'. Keynes' notion deals with this maintained hypothesis itself, more precisely with the existence of a meaningful conditional probability distribution.
6. The principle dates back to Aristotle, cited in Derkse (1992, p. 13): 'Nor can the factors of Nature be unlimited, or Nature could not be made the objects of knowledge.'

7. The Cambridge tradition of philosophy of G. E. Moore, in which Keynes operated, was preoccupied with analysing the meaning of words in terms of their simplicity or complexity. A quintessential example is the notion 'good', which was thought to be a simple, non-natural property to be grasped by intuition. This means that 'good' cannot be reduced to other terms but can only be understood intuitively (see Davis, 1994). This approach in Moore's theory of ethics was extended to a theory of induction.
8. Stigum (1990, p. 567; see also p. 545) provides a formal version of the principle of limited independent variety. He argues that it is needed in conjunction with the principle of the uniformity of nature. My view is that it is a weaker version of the principle of uniformity of nature.
9. In this sense, the inductive hypothesis resembles a Kantian synthetic *a priori* proposition.
10. Carabelli (1985) argues that Keynes' intuitionism makes his probability theory anti-rationalistic, and his logicism makes his theory anti-empirical. I do not concur in these views. It is better to interpret Keynes' theory as one of bounded rationality. The rules of probabilistic inference are logical in kind, but applications may be empirical. The judgement about the validity of the application itself is not empirical, but logical. The decision to accept the validity may be regarded as boundedly rational.
11. Keynes' discussion of this contest in ([1936] CW VII, p. 156) serves a slightly different purpose, by showing how expectations are interdependent and, therefore, undetermined: 'It is not a case of choosing those which, to the best of one's judgement, are really the prettiest, nor even those which average opinion genuinely thinks the prettiest. We have reached the third degree where we devote our intelligences to anticipating what average opinion expects the average opinion to be. And there are some, I believe, who practise the fourth, fifth and higher degrees.'
12. Compare this with the later research on consumer demand, discussed in chapter 8. A complete shift in the emphasis of testing can be observed. In the more recent examples of consumer demand, the focus is on significance testing. The quantitative importance of deviations of parameters from the desired values is hardly every considered.
13. Savage was a student when Carnap taught at Chicago. They had occasional talks, but primarily at the Princeton Institute for Advanced Studies (see Carnap and Jeffrey, 1971, p. 3). Another missed chance of direct impact was Herbert Simon, who studied with Carnap. Carnap's influence on Simon's theory of inference is small (see Langley *et al.*, 1987).
14. See von Neumann and Morgenstern ([1944] 1947). They use a frequency interpretation of probability, but a footnote suggests modelling preferences and probability jointly (p. 19). This is done by Savage ([1954] 1972). This is not just a coincidence: Savage was one of von Neumann's wartime assistants.
15. There was occasional contact between Savage and the econometricians. Savage ([1954] 1972, p. xi) thanks Tjalling Koopmans in his preface. Savage worked together with Milton Friedman, not a Cowles member.

Friedman writes (in private correspondence) that, although Savage had many interests, they were more in the area of medicine than in econom(etr)ics. Unlike Friedman, Savage did not actively participate in the activities of Cowles. See also Hildreth (1986, p. 109, n. 11).

16. For a discussion, see Barnett (1973), Cox and Hinkley (1974) and Berger ([1980] 1985).
17. An excellent survey of such non-informative priors can be found in Zellner (1971, pp. 41–54).
18. During the Second World War, Jeffreys taught his probability course to one of the leading postwar econometricians, Sargan. Sargan experimented with Bayesian statistics afterwards, but he gave up fairly soon because of the objectivity problem of Bayesian statistics. One of the two other students in the class was the Bayesian probability theorist Lindley. Jeffreys' classes were not popular because of his unattractive style of teaching, basically consisting of reading excerpts from his book in a monotone (Sargan, private conversation, 1 December 1992).

5 The theory of simplicity

> Is it not natural to begin any attempt at analysing the economic mechanism by making the simplest assumption compatible with general theory?
>
> Jan Tinbergen (1940)

1 Introduction[1]

'The march of science is towards unity and simplicity', Poincaré ([1903] 1952, p. 173) wrote. In econometrics, this march threatens to become lost in a maze of specification uncertainty. There are many different ways to deal with specification uncertainty. Data-mining invalidates traditional approaches to probabilistic inference, like Fisher's and Neyman–Pearson's methodology. A theory of simplicity is helpful to get around this problem.

Fisher (1922, p. 311) defined the object of statistics as 'the reduction of data'. The statistical representation of the data should contain the 'relevant information' contained in the original data. The decision, about where to draw the line when relevant information is lost is not trivial: Fisher's own methodology is based on significance tests with conventional size (usually, 0.05). At least, Fisher had relatively sound *a priori* guidelines for the specification of the statistical model. This luxury is lacking in econometrics. Fisher could rely on randomization, to justify his 'hypothetical infinite population' of which the actual data might be regarded as constituting the random sample. Chapter 6 shows that economic metaphors referring to this hypothetical infinite population are unwarranted.

Another effort to justify Fisher's reduction can be found in structural econometrics, where the model is based on a straitjacket imposed by economic theory. This provides the *a priori* specification of the econometric model. In practice, this methodology does not work, as chapter 7 illustrates. Three other econometric methodologies, also discussed in chapter 7, are (model-)reductionism (Hendry, 1993), sensitivity analysis (Leamer, 1978) and profligate Vector Autoregressive (VAR) modelling (Sims, 1980). They share, as their starting point, a general, high-dimensional model. The *reductionist* approach attacks specification uncertainty through a general to specific 'simplification search' by means of conditioning and marginalizing, in which a sequence of (possibly) asymptotically independent significance tests is used. However, the notion of simplicity is not explicitly formalized. Instead, a conventionally chosen significance level is used. The *sensitivity* approach does not have serious interest in simplicity. Still, Leamer prefers to model from 'general' to 'specific', but he does not rely on one specific favourite model as 'the' result of inference. He only trusts specific outcomes if they are robust for acceptable changes in the specification. The *VAR* approach, finally, uses information criteria in order to obtain models with 'optimal' lag lengths, but no effort is made to use such criteria for imposing additional zero restrictions on the parameters of the model. As a result, VAR models are often criticized for being 'over-parameterized', or insufficiently parsimonious.

Simplicity is appealing though hardly operational. In this chapter, an effort is made to formalize simplicity and to show how it relates to scientific inference and econometric modelling. In particular, use will be made of insights in algorithmic information theory, the theory of inductive reasoning and Kolmogorov's complexity theory. It will be argued that simplicity is a vital element in a theory of scientific inference. A definition of simplicity is suggested, as are the optimum conditions for the desired degree of simplicity. Understanding simplicity is seen to be important for specifying hypotheses and selecting models.

The chapter is organized as follows. Section 2 discusses the background and formalization of simplicity with regard to *a priori* probability and information theory. Section 3 deals with its meaning for scientific inference in general and its relation to induction, the Duhem–Quine thesis, the purpose of a model and bounded rationality. Some concluding comments are given in section 4. An important issue remains how the theory of simplicity may clarify problems of specification uncertainty and econometric modelling. This issue is considered in chapter 7.

2 Formalizing simplicity

2.1 Background: Ockham's razor and the law of parsimony

A famous maxim of both philosophy, in general, and scientific inference, in particular, is Ockham's razor: '*Entia non sunt multiplicanda praeter necessitatem*' (entities are not to be multiplied beyond necessity). This maxim is attributed to William of Ockham (*c.* 1285–1349).[2] Different versions of the principle circulate,[3] and even the spelling of Ockham's name varies – the razor has been used to shave it to Occam. Ockham's razor is one of the canons of mediaeval scholasticism (see e.g. Pearson, [1892] 1911, p. 393, but see Thorburn, 1918, for a diverging view). Related canons, cited by Pearson ([1892] 1911, p. 393), are *principia non sunt cumulanda*, and a statement that can be found in the writings of Ockham as well as those of his teacher Duns Scotus (*c.* 1266–1308), *frustra fit per plura, quod potest fieri per pauciora* (it is in vain to do by many what can be done by fewer). This is known as the law of parsimony (or parcimony).

Whatever the antecedents of the razor, as a methodological principle it is useful but vague. Other things equal, a more elaborate model cannot fit the data worse than a specific (restricted) version of it. Ockham's razor suggests deleting those extensions of a model that are irrelevant to the aim of the model, examples of aims being description and prediction. It is a recurring theme in the writings of early positivists, such as Mach, Poincaré and Pearson (see e.g. Pearson, [1892] 1911). If two hypotheses H_i and H_j describe data D equally well, i.e. $P(D|H_i) = P(D|H_j)$, the principle implies that the simpler of the two should be prefered.

But what is the motivation? Some efforts have been made to base simplicity on metaphysical grounds. An example is the view of Sir Isaac Newton (1642–1725) that '*natura enim simplex est et rerum causis superfluis non luxuriat*' (*Principia Mathematica*, 1687, cited in Pearson, [1892] 1911, p. 92).[4] Newton holds that nature is simple, and no more causes should be admitted than 'such as are both true and sufficient' to explain the phenomena (Newton, in: Cohen and Westfall, 1995, p. 116). The view that nature is simple is shared by many physicists. For example, it can be found in a paper on the principles of scientific inquiry by Wrinch and Jeffreys (1921, p. 380): 'The existence of simple laws is, then, apparently, to be regarded as a quality of nature.' Another modern version of Newton's view is given by the physicist Richard Feynman, who wonders about the ability to guess nature's laws. His answer is: 'I think it is because nature has a simplicity and therefore a great beauty' (Feynman, 1965, p. 173). One might object that this view of the physicist is metaphysical speculation, not based on facts. A diametrical speculation is expressed by Leamer and Hendry, who agree that 'Nature is complex and Man is simple' (Hendry, Leamer and Poirier, 1990, p. 185).

Ockham's razor does not imply that nature is either simple or complex, but only suggests that simplicity is a sound device for inference. It serves as a rule of methodology, not as a metaphysical dogma (Thorburn, 1918, p. 352). This interpretation concurs with the writings of positivists such as Mach, Pearson and Poincaré (see [1903] 1952, p. 146). The latter notes that 'any fact can be generalized in an infinite number of ways, and it is a question of choice. The choice can only be guided by considerations of simplicity.' The example he discusses is the interpolation of a number of points in a plane. Why not describe the points using a curve with the 'most capricious zigzags? It is because we know beforehand, or think we know, that the law we have to express cannot be so complicated as all that.' He continues, 'To sum up, in most cases every law is held to be simple until the contrary is proved.' This is one of the first explicit discussions of the importance of simplicity in the probabilistic approach to scientific inference. A pragmatic argument is added by Peirce (1955, p. 156):

> Hypothesis in the sense of the more facile and natural, the one that instinct suggests, that must be preferred; for the reason that, unless man have a natural bent in accordance with nature's, he has no chance of understanding nature at all.

Boltzmann ([1905], cited in Watkins, 1984, p. 107), who argued that an hypothesis is false if a simpler hypothesis is superior in representing the data, pointed to the problem that simplicity is not operational. Simplicity, he argues, is largely a subjective criterion. Therefore, 'science loses its stamp of uniformity'. When, for example, are new elements in a theory redundant? To answer this question, it is necessary to face the trade-off between simplicity and (descriptive) accuracy.

This trade-off has been studied by some of the founders of probability theory. An early attempt was made by Leibniz.[5] He argues that an hypothesis is more probable than another in proportion to its simplicity (economy of assumptions) and its power (number of phenomena that can be explained by the hypothesis; see Keynes, [1921] CW VIII, p. 303; Cohen, 1989, p. 27). In his discussion of Leibniz, Keynes does not elaborate on the possibility of formalizing the trade-off. Keynes' own views on parsimony are to be found in the principle of limited independent variety, which may justify the use of non-zero *a priori* probabilities in probabilistic inference. It is not a methodological rule to trade-off simplicity and goodness of fit. Ramsey (cited in Edwards, [1972] 1992, p. 248), mentioned the trade-off and suggested making use of maximum likelihood:

> In choosing a system we have to compromise between two principles: subject always to the proviso that the system must not contradict any facts we know, we choose (other things being equal) the simplest system, and (other things being equal) we choose the system which gives the highest chance to the facts we have observed. This last is Fisher's 'Principle of Maximum Likelihood', and gives the only method of verifying a system of chances.

Kemeny (1953) discusses the same issue. He defines compatibility of a theory with the observations by the condition that the observations must lie within a 99% confidence interval (p. 398). He (p. 397) suggests adopting a rule of inference, which is to select the simplest hypothesis compatible with the observed data. This is his third rule, which has been named 'Kemeny's Rule' by Li and Vitányi (1992). Although intuitively appealing, this is rather *ad hoc*. Kemeny (1953, p. 408) remarks that it would be of interest to find a criterion combining an optimum of simplicity and compatibility based on first principles. The optimum trade-off is not presented.

Not surprisingly it is, therefore, frequently argued that Ockham's razor is arbitrary, even as a methodological rule (e.g. Friedman, 1953, p. 13 n.). Friedman argues that predictive adequacy is the first criterion to judge a theory, simplicity the second, while realism of assumptions is of little interest. However, he does not define those terms. Moreover, the view that prediction is paramount in the sense that it provides greater support for a theory than does, say, the testing of auxiliary hypotheses, is disputable (Howson, 1988a) and should be justified. Principles of simplicity that are used in econometrics (such as Theil's $\bar{R}^2$, or the information criteria of Akaike and Schwarz) are often discussed as being useful but, again, *ad hoc* (see e.g. Judge *et al.*, [1980] 1985, p. 888; Amemiya, 1985, p. 55).

2.2 Simplicity and a priori *probability*

Wrinch and Jeffreys (1921, p. 389) provide an explicit discussion on the relevance of simplicity in scientific inference. They note that 'it will never be possible to attach appreciable probability to an inference if it is assumed that all laws of an infinite class, such as all relations involving only analytic functions, are equally probable *a priori*'. In order to enable inference, they invoke two premises (p. 386):

- Every law of physics is expressible as a differential equation of finite order and degree, with rational coefficients; and
- the simpler the law, the greater is its prior probability.

The first premise is like Russell's (1927, p. 122) currently outdated claim that scientific laws of physics can only be expressed as differential equations. Once laws are expressed in differential form, they can be ordered, beginning with differential equations of low order and degree, involving no numerical constants other than small integers and fractions with small numerators and denominators (Wrinch and Jeffreys, 1921, p. 390). Although highly informal, this is the first attempt to operationalize the notion of simplicity. Once this is done, the ordering is used in a probability ranking, using the second premise. In practice, Wrinch and

Jeffreys (p. 386) argue, simplicity 'is a quality easily recognizable when present'.

In his later writings, Jeffreys elaborated his views on simplicity. Jeffreys ([1939] 1961) argues that there is good reason to give a simple hypothesis a higher prior probability than a complex hypothesis, in particular because simple hypotheses tend to yield better predictions than complex hypotheses: they have superior inductive qualities. The argument does not rest on a view that nature is simple. Simple hypotheses are favoured *a priori* on grounds of degrees of rational belief (pp. 4–5). As a result, hypotheses of interest can be analysed using strictly positive prior probabilities. The argument is known as the Jeffreys–Wrinch Simplicity Postulate (SP) (Jeffreys, [1931] 1957, p. 36):

Simplicity Postulate (SP). The set of all possible forms of scientific laws is finite or enumerable, and their initial probabilities form the terms of a convergent series of sum 1.

Jeffreys ([1939] 1961, p. 47) summarizes his view as 'the simpler laws have the greater prior probability'. The simplicity postulate in the crude operational form given by Jeffreys (p. 47) is to attach prior probability 2^{-c} to the disjunction of laws of complexity c, where c is measured by the sum of the

- order,
- the degree, and
- the absolute values

of the coefficients of scientific laws, expressed as differential equations. Jeffreys proposes assigning a uniform prior probability distribution to laws with an equal degree of complexity. There are two problems with Jeffreys' approach. The first serious problem is that his measure of complexity remains arbitrary, although it is more explicit than the one presented in the paper with Wrinch.[6] Furthermore, as he acknowledges, it rules out measures of complexity of hypotheses which cannot be formulated as differential equations. Jeffreys provides neither a formal justification for the simplicity postulate nor for the 2^{-c} rule, and acknowledges that the prior probabilities assigned to scientific laws may not be sufficiently precise (pp. 48–9). The simplicity postulate is an heuristic principle, with one feature that has to be retained in any event, Jeffreys (p. 48) argues:

> If a high probability is ever to be attached to a general law, without needing a change in the form of the law for every new observation, some principle that arranges laws in an order of decreasing initial probability with increasing number of adjustable parameters is essential.

A justification for a modified SP, and suggestions for obtaining sufficiently precise prior probabilities needed to implement the modified SP, are given below, after discussing the second problem of the simplicity postulate. This problem, which seems an obvious and devastating blow at first sight, is discussed in Forster (1995, p. 407). He claims that the simple hypothesis, H_1 ($s = a$), logically entails the more complex one, H_2 ($s = a + ut$) for $u = 0$. This, he argues, is 'very embarrassing' to Bayesians. Strong propositions like H_1 cannot have higher probability than weaker ones, unless the probability axioms are violated. To Forster, this is a 'decisive' refutation of the simplicity postulate.

A first rebuttal is that there is a difference between $s = a$ and $s = a + 0t$, due to the different description lengths of the two propositions and the implicit information on the vector t which is part of the second expression. This difference has been illuminated by probabilistic information theory, discussed below. Second, there is nothing wrong if a subjectivist assigns a probability $P(H_2|u = 0) > \frac{1}{2}$ and $P(H_2|u \neq 0) < \frac{1}{2}$ if the variable t has no prior justification (i.e. is *ad hoc*). This is a reason why many economists reject demand equations which are non-homogenous in income and prices, or disequilibrium dynamics in neoclassical-Keynesian synthesis models. Although their *fit* cannot be worse than restricted (parsimonious) models, their inductive performance (as, for example, predictive power) may be worse. Forster's embarrassment only applies to Bayesians without inductive aims, and those are rare. Third, the two propositions which Forster compares are not the type of propositions which are at stake in statistical inference. The error term is missing in both of them, and they will have different properties. It is not a logical truth that $s = a + \epsilon$ entails $s = a + ut + \epsilon^*$, hence, the argument of Forster does not hold.

2.3 *Simplicity and information theory*

A justification for the trade-off between descriptive accuracy and parsimony can be found in the literature on algorithmic information theory. This approach to inference starts from the idea that knowledge, or information, can be represented by strings of zeros and ones. Such strings of binary digits (bits), denoted by $\{0, 1\}^*$, can store models (hypotheses, theories) as well as experience (data). A theory is a computer program that computes binary sequences which mimic the binary representation of the observed data. The goal is to design a computer program that, on the one hand, is as simple as possible, whilst on the other hand is compatible with the observed data.

The simplicity of an hypothesis is related to the length of the string describing the hypothesis, and this length is related to a probability measure. A problem is that this length depends on the language in which the hypothesis is expressed. It can be shown, however, that a minimum code length for an hypothesis does exist (Rissanen, 1983), but this code length is not computable. The problem then is to find a good approximation to this minimal code length. Hence, a formal justification of Ockham's razor is feasible, although the implementation of the resulting principle is arbitrary. Heuristic arguments will be presented to substantiate this claim (proofs can be found in Li and Vitányi, 1992).

According to Bayes' theorem, the posterior probability for an hypothesis H_i given data D is proportional to the prior times the likelihood. It will be assumed that the hypotheses under consideration are at most countably infinite.[7] The goal is to maximize the posterior probability $P(H_i|D)$: the more elaborate is H_i, the better will the hypothesis fit the data and the larger is the likelihood, $P(D|H_i)$. Simultaneously, however, according to the Simplicity Postulate, the prior probability, $P(H_i)$ declines with the increasing complexity of the hypothesis.

The more elaborate is an hypothesis, the greater is the number of bits required for its binary representation. Rewriting Bayes' theorem in negative logarithms (where $\log(\cdot)$ denotes logarithms with base 2, which is a useful convention in a binary context), yields:

$$-\log P(H_i|D) \propto -\log P(D|H_i) - \log P(H_i). \tag{1}$$

Maximizing the posterior probability (selecting the hypothesis that has the highest support of the data) is equivalent to minimizing this expression.

According to algorithmic information theory, $\log P(H_i)$ is related to the descriptive complexity K of the hypothesis H_i and $\log P(D|H_i)$ is related to the 'self-information' of the data given H_i. Roughly speaking, K measures the minimum number of bits required to encode a proposition. The relation between complexity and a probability distribution has been established by Solomonoff (1964a) and Levin (see Rissanen, 1983). Solomonoff discusses a general problem of induction: extrapolating a sequence of symbols drawn from some finite alphabet. Given such a sequence, denoted by S, what is the probability that it will be followed by a sequence a? Hence, the problem is to calculate:

$$P(Sa|S) = \frac{P(S|Sa)P(Sa)}{P(S)}. \tag{2}$$

One of the examples, presented in Solomonoff (1964b), is the extrapolation of a Bernoulli sequence (a string of zeros and ones). The analytic approach starts with (1). The prior probability in (1) is obtained by examining how the strings of symbols might be produced by a 'Universal Turing Machine' (UTM).[8] The crucial insight is that strings with short and/or numerous descriptions (program codes that yield the relevant string as output) are assigned high *a priori* probability. Solomonoff gives intuitive as well as quasi-formal motivations for this assignment (a purely formal motivation can be found in Li and Vitányi, 1992). The intuitive motivation is based on two arguments. First, the resulting approach of inference is explicitly related to Ockham's razor, 'one interpretation of which is that the more "simple" or "economical" of several hypotheses is the more likely' (Solomonoff, 1964a, p. 3). Second, strings with many possible descriptions yield high *a priori* probability because of the 'feeling that if an occurrence has many possible causes, then it is more likely' (p. 8). The quasi-formal motivation for Solomonoff's argument to relate the prior probability to simplicity comes from coding theory, from which a measure is derived about the likelihood that a will follow S. This measure obeys 'most (if not all) of the qualities desired in an explication of Carnap's probability$_1$'. Later developments of algorithmic learning theory have improved upon Solomonoff's pioneering work, without losing the basic insights about the relation between simplicity and probability. The result is known today as the Solomonoff–Levin distribution with respect to a UTM, which will be denoted by $P_{UTM}(H_i)$. In case of (2), it can be shown that $2^{-K(Sa)} \leq P_{UTM}(Sa)$ (e.g. Li and Vitányi, 1992). It turns out that Jeffreys' prior probability distribution, based on the Jeffreys–Wrinch Simplicity Postulate, is an approximation to the ideal (or 'universal', as Rissanen, 1983 calls it) prior probability distribution.[9] The Solomonoff–Levin distribution is not based on subjective Bayesian betting, but on 'objective' features of the problem of interest. Therefore, it may be regarded as an element in a 'necessarist' theory of Bayesian inference.

A weakness of the sketched approach is that the resulting probability distribution is not effectively computable in the sense of Turing. This is due to the halting problem (see footnote 8). There are limitations on the ability of the Turing machine to foresee whether a particular program code c will provide any output or will run forever, entering infinite loops. Therefore, a computable approximation to $P_{UTM}(Sa)$ must be obtained. Solomonoff (1964b) contains a number of examples in which his insights are applied, showing how well some pragmatic approximations perform. Rissanen (1983) and Li and Vitányi (1992) discuss a number of approximations to Solomonoff's ideal inductive method, including maximum likelihood, maximum entropy and Minimum Description Length (MDL, also known as global maximum likelihood). Among the most interesting computable approximations is Rissanen's (1983) MDL principle.

2.4 Minimum description length (MDL) and maximum likelihood

Rissanen (1983) shows the links between the MDL principle and maximum likelihood. The rather obvious link to the simplicity postulate will also be presented. First consider MDL. Recall the code length of a model, parameterized by θ, and the data, represented by x. The length $l(x, \theta)$ consists of the self-information plus the model code, and is given by

$$l(x, \theta) = -\log P(x|\theta) + l(\theta). \tag{3}$$

The formula is related to Bayes' theorem, written in logarithmic form. If $2^{-l(\theta)}$ is written as $Q(\theta)$ (which can be shown to satisfy the conditions for a probability distribution, see Rissanen, 1983), then

$$2^{-l(x,\theta)} = P(x|\theta)Q(\theta). \tag{4}$$

The parameter for the code length, θ, is an integer and must have a prior probability distribution on all its possible values, i.e. on all integers. To encode an integer scalar θ, approximately $\log n(\theta)$ bits are required (where the integer $n(\theta)$ is the binary representation of θ divided by its precision). The prior probability $Q(n) = 2^{-\log(n)} = 1/n$ seems a natural prior probability distribution (namely, the uniform distribution proposed by Jeffreys), but this is not a proper one as it does not sum to 1. Another (related) problem is that the computer must be able to separate different strings, without dropping the binary character of encoding. The prefix property is needed to solve this difficulty, where a prefix code is a program or the binary representation of a model. This implies that a number greater than $\log n$ bits is required to encode the integer n. The exact number of minimally required bits of the prefix code is not computable. The required code length can be approximated by $\log^*(n) \equiv \log(n) + \log\log(n) + \ldots$ (the finite sum involves only the non-negative terms; see Rissanen, 1983, p. 424).[10] The proper prior probability that results is:

$$Q(n) = \frac{1}{n} \times \frac{1}{\log(n)} \times \ldots \times \ldots \frac{1}{\log\ldots\log(n)} \times \frac{1}{c}, \tag{5}$$

where c is a normalization constraint. The first (and dominant) fraction corresponds to Jeffreys' non-informative prior.

The universal prior probability distribution is used to construct an 'objective' method of inference, which is sometimes called global maximum likelihood, or the principle of minimum description length (for derivations, see Rissanen, 1983, 1987). Let $P(x|\theta)$ be the likelihood of the n observations of data x, parameterized by the k-dimensional parameter vector θ. $I(\theta) = M(\theta)/n$, where $M(\theta)$ is the Hessian of $-\log P(x|\theta)$ (the information matrix) evaluated at the maximum likelihood estimate. Then the minimal description length MDL of the data encoded with help of the theory, and the theory itself, is (up to a term of order $(\log k)/n$).

$$MDL = \min_{\theta,k}\left[-\log P(x|\theta) + \frac{k}{2}\log\frac{2\pi en}{k} + k\log\|\theta\|\right]. \tag{6}$$

The third term containing the norm of θ, evaluated at the optimum, makes the criterion invariant to all non-singular linear transformations. The first two terms in the minimized expression correspond to Schwarz's (1978) Bayesian Information Criterion, which is:

$$SIC = \min_{\theta,k}\left[-\ln P(x|\theta) + \frac{k}{2}\ln(n)\right]. \tag{7}$$

A related information criterion which is slightly simpler and was derived earlier, is the Akaike Information Criterion (AIC):

$$AIC = \min_{\theta,k}[-\ln P(x|\theta) + k]. \tag{8}$$

Minimizing AIC has also been called the Principle of Parsimony (Sawa, 1978). Compared to the MDL principle and SIC, AIC has a few problems. One objection is that it fails to give a consistent estimate of k (Schwarz, 1978). Another objection relates to the dependence on some 'true' distribution which relies on an assumption that the pseudo-true model is nearly true (Sawa, 1978, p. 1277; see Zellner, 1978, for a comparison of AIC with Bayesian posterior odds; Forster, 2000, presents a defence of AIC).

Rissanen (1983, p. 428) remarks that applying the MDL principle, while retaining only the first two terms of (6), has been used successfully for the analysis of autoregressive (AR) and autoregressive moving average (ARMA) models – a context where the other information criteria are popular as well. Furthermore, Rissanen writes,

In contingency tables, the criterion, measuring the total amount of information, offers a perhaps speculative but nonetheless intriguing possibility to discover automatically inherent links as 'laws of nature' in experimentally collected data. In the usual analyses such links had to be first proposed by humans for a statistical verification or rejection.

The somewhat naive goal of the algorithmic information-theoretical approach is the automation of science: making the *machina ratiocinatrix*, the reasoning machine (Wiener, [1948] 1961, pp. 12, 125) operational.[11] However, the smaller the number of observations, the more arbitrary will be the computable approximation to the 'universal' prior probability distribution. The automation of science may be the dream of Asymptopia, but is unlikely to be the future of econometrics.

Indeed, it is doubtful whether automated methods would work for discovery in economics. Econometric inference is different from the inferences that have been investigated by computer scientists and workers in artificial intelligence. The complexity of economic problems exceeds the complexity of the problems varying from inferring Kepler's laws to recognizing on-line written Chinese characters. The analogy between on-line handwritten character learning and economic inference becomes more realistic if we imagine a writer who knows that a computer is trying to read his handwriting. Then he might change his style of writing, for fun, or to deceive the computer, or because of learning feedback.

Finally, compare Rissanen's prior (5) to Jeffreys' Simplicity Postulate combined with inference based on an improper uninformative prior. Jeffreys starts from encoding an hypothesis H to an integer belonging to $\mathbb{N}^+ = \{1, 2, \ldots\}$. Examples of efficient coding can be found in Solomonoff (1964b). Assign prior probability $1/i$ to integer i. Although this results in an improper distribution, as $\Sigma 1/i$ diverges from 1 as $i \to \infty$, the procedure may work well if only a small number of hypotheses is being considered. The prior is an (improper) approximation to the universal prior. To see this, note that $1/i = 1^{-\log(i)}$, whereas Rissanen's (proper) approximation to the universal prior is $2^{-\log^*(i)}$.

Obviously, the method of ML is another special case of global maximum likelihood. Given an hypothesis H that defines the parameters θ of a model, and given the data x, the goal is to maximize $\ln P(x|\theta)$. If the total code length is given by $l(x, \theta) = -\log P(x|\theta) + l(\theta)$ then ML amounts to minimizing the code length for a fixed parameterization θ, as only $\log P(x|\theta)$ is evaluated under the 'axiom of correct specification'. In econometric practice, maximum likelihood is rarely applied without serious pre-testing (guided by vague preferences for simplicity). The attractive feature of global maximum likelihood is that it provides a well-founded criterion for the trade-off between goodness of fit and parsimony. The modified simplicity postulate is to minimize the total code length needed to describe the data with respect to a particular model. Unlike Jeffreys' original SP, the modified postulate is well defined. An unavoidable arbitrariness in implementing this postulate remains, but the amount of arbitrariness can be made precise and is easy to understand.

3 Simplicity and scientific inference

3.1 Simplicity and induction

Understanding simplicity is crucial for appraising methodological debates. Surprisingly, the methodological literature in economics ignored this issue until quite recently. In the philosophy of science literature, simplicity has received a stronger interest. An important contribution is Popper ([1935] 1968, p. 385) who argues that simple theories should have low (or zero) prior probability: 'simplicity, or paucity of parameters, is linked with, and tends to increase with, improbability rather than probability'. He concludes that all theories of interest (in particular, theories that are simple and easy to falsify, i.e. theories with 'high empirical content') are simple to such an extent that they even have zero prior probability. If Popper is correct, then the Bayesian (inductive) approach to inference cannot be sustained, as all posterior probabilities of non-tautological hypotheses would be zero. This is why Popper argues in favour of falsificationism: if a degree of support for scientific propositions cannot be established, at least false propositions can be eliminated. Science thereby progresses by a sequence of conjectures and refutations.

Such arguments have been criticized by the Bayesian philosopher Howson (1988a), who sees no role for simplicity in epistemology. Howson rejects the simplicity postulate, arguing that it is as arbitrary as Popper's opposing argument (see also Howson and Urbach, 1989, p. 292). Howson and Urbach are not the only Bayesians who reject the importance of simplicity. For example, Leamer (1978, p. 203) claims to be agnostic. Leamer (in Hendry, Leamer and Poirier, 1990, p. 184) argues that simplicity may be helpful for the purpose of communication, but not for inference. I believe this is a serious mistake: simplicity is essential for a theory of inference, in particular Bayesian inference.

Keynes and Jeffreys rightly emphasized that, in order to use probabilistic reasoning for scientific inference, one has to map probabilities *a priori* to probabilities *a posteriori*. The probability *a priori* of an hypothesis should not be negligible, otherwise the posterior probability will be zero. If there is no prior information on the merits of alternative hypotheses, it is necessary to make a non-informative prior probability assignment. The uniform distribution has been used for this purpose. If the number of alternative hypotheses is unbounded, each hypothesis will have a prior probability of zero. Hence, the posterior probability of scientific hypotheses will be zero as well, so that inductive inference is not possible. Keynes and Jeffreys suggested different approaches to solve this problem of induction, with Keynes invoking the principle of limited independent variety (discussed in chapter 4, section 2.2) and Jeffreys the simplicity postulate. The latter relates directly to Ockham's razor. Jeffreys ([1939] 1961, p. 342) interprets the statement 'entities are not to be multiplied without necessity' as follows:

Variation is random until the contrary is shown; and new parameters in laws, when they are suggested, must be tested one at a time unless there is specific reason to the contrary.

In the preface to the third edition of his *Theory of Probability*, Jeffreys (p. viii) notes that the implications of this principle are contrary to 'the nature of induction as understood by philosophers' (presumably, Popper is implicated). The important point, which can already be found in Pearson ([1892] 1911), is that one starts with the hypothesis that variation in the data is random, and then gradually elaborates upon a model to describe the data in order to improve upon the approximation. Furthermore, Jeffreys ([1939] 1961, p. 129) argues that

Any clearly stated law has a positive prior probability, and therefore an appreciable posterior probability until there is definite evidence against it.

The simplest hypothesis is that variation is random until the contrary is shown, the onus of the proof resting on the advocate of the more complicated hypothesis (pp. 342–3).

It is now possible to compare Jeffreys' approach with Popper's. Although it is acknowledged that Jeffreys' simplicity postulate is made operational in an arbitrary way, the heuristic argument is sound. More recent advances, in particular due to Rissanen (1983), provide a non-arbitrary operational form of the postulate. On the other hand, Popper's critique of assigning positive probability *a priori* to simple propositions is self-defeating. The argument is based on two points: first, the idea that the number of possible propositions is infinite; second, that the empirical content of simple propositions is high, so that the chances of falsification are also high. On the first point, if the number of conjectures is indeed infinite, the chance that a sequence of conjectures and refutations will come to an end by hitting the truth is zero. A methodology based on gradual approximation does not suffer from such a weakness. The second point underlying Popper's argument is due to confusing $P(H)$ and $P(D)$.

3.2 Simplicity and the Duhem–Quine thesis

The Duhem–Quine thesis of testing theories suggests that a negative test result cannot disprove a theory: the rejection affects one (or more) element(s) of a whole test system, but does not indicate which element is invalidated by the test. If a prediction of a scientific theory is shown to be wrong, one can adjust any part of the test system (sometimes called 'web' or 'cluster'), and not just the hypothesis of interest. The simplicity postulate reduces the impact of the Duhem–Quine thesis, although its logical status is not altered.

According to the modified simplicity postulate, simple models have stronger *a priori* predictive power (in the sense of minimizing a mean squared prediction error) than complex models, unless specific prior information is available to justify complexity. Adding *ad hoc* auxiliary hypotheses to a model, and using them for modified predictions based on the revised test system, will reduce the *a priori* predictive power of the model. The reason is that '*ad hockery*' is unrelated to the other theoretical notions that are used to establish a model. Hence, *ad hoc* alterations of theory need a larger additional code (bits) and, therefore, increase the total descriptive length. Negative test results should lead to efforts to adapt the model as coherently as possible, minimizing '*ad hockery*' (hence, additional code length). This view receives support from Quine's (1987, p. 142) suggestion to introduce the 'maxim of minimum mutilation' in responding to a test result: 'disturb science as little as possible, other things equal'.

This is important for discussions of data-mining. The data-miner has absolute freedom to make any changes in the specification of a model. As the best (perfect) fit that can be obtained will be the one resulting from a model with as many parameters as there are observations, the data-miner tends to make an implicit trade-off between fit and simplicity. For example, the SAS statistical package provides statistical routines where, given a set of data for particular variables, the linear model yielding a maximum $\bar{R}^2$ is selected. The stepwise regression algorithm is a simplified version of this automated modelling procedure. A major problem with such routines is the *ad hoc* nature of the resulting models. Simplicity and fit are considered and, if description of a particular set of data is the only purpose of interest, then the resulting model may be satisfactory. Similarly, if a Box–Jenkins (time-series) model is chosen by optimizing a particular information criterion, a satisfactory model for describing the given data may be obtained. However, such models are idiosyncratic and do not share general results with other models. Hence, investigators with broader aims (such as scientific inference, induction) are unlikely to opt for mechanical modelling using, for example, stepwise algorithms. The role of economic theory is to highlight general characteristics of different sets of data, in which case economic theory is a simplifying device.[12] This can be formulated in terms of the modified SP. If an investigator analyses consumption in both the USA and the UK, applying Box–Jenkins techniques, he is likely to yield two statistically adequate models with very different specifications and, hence, a larger number of bits will be required to describe them. A theoretical straitjacket is likely to yield models that are slightly worse in terms of goodness of fit, but with greater simplicity in terms of the number of bits, due to the similar specifications involved. This analysis of the Duhem–Quine problem avoids the problems that result from a Lakatosian interpretation of scientific research programmes (advocated by Cross, 1982). It provides a justification for structural econometrics (see chapter 7). This justification is not that such econometrics makes economic theory falsifiable (because of the Duhem–Quine thesis it does not), but because it yields relative simplicity (in particular if exogeneity can be exploited) and non-idiosyncratic models. The simplicity postulate penalizes '*ad hockery*' or, in Popperian terms, immunizing stratagems. If the latter is the sound core of Popperian methodology (as Caldwell, 1991, claims), than Popper's opponent Jeffreys rather than Popper himself has provided an acceptable justification.

3.3 The purpose of a model

Has simplicity a virtue independent of the goal of inference? Rissanen's hope for the automation of science suggests that simplicity is an objective characteristic independent of subjective judgements of an investigator or independent of the context of modelling. Rissanen (1987, p. 96) argues that the introduction of subjective judgements in inferential problems makes the resulting inferences 'strictly speaking unscientific'. However, objectivity is a deceptive notion (this is particularly true for economic inference where data information is limited and theory-loaded and, as a result, value judgements are inevitable). The modified simplicity postulate may help, however, to yield greater agreement on how to trade off parsimony and descriptive accuracy. Algorithmic information theory shows why certain constraints to objective knowledge may exist, but it does not lead to the conclusion that subjective judgements are unscientific.

Subjective judgements matter, and so does the context of inference. Prediction is different from regulation and control, for example. Relatively complex models may be needed if policy intervention is the purpose of modelling (this belief stimulated large scale macro-econometric modelling in the '60s and '70s, and subsequently, their complex successors with elaborate micro-foundations). Nevertheless, this does not violate the (modified) simplicity postulate. Given the goal of inference, the rule should be to select the set of models that is able to meet the goal, so that the postulate is used in order to choose the appropriate model.

It might be objected that the minimum description length principle does not pay sufficient attention to the purpose of the modeller. If, from the MDL perspective, Einstein's relativity theory is superior to classical mechanics in describing red shifts in the spectra emitted by stars, should it be used to predict the position at time t of a football that has been set in motion at $t-1$? The answer is negative: if the input data are macroscopic, then it is likely that inductive inference related to the position of the football will yield superior results using classical mechanics. The MDL principle would suggest that the extra bits to describe the relativity part of the more complex model do not improve the predictions and should, therefore, be disregarded. No physicist would be likely to consider starting from the relativity model and test downwards in this case.

If the goal of inference is to make money, for example, by predicting stock prices, the criterion is not to maximize simplicity but to maximize dollars. The example in Rissanen (1987) is interesting in this respect: applying the MDL principle generates the random walk model of stock prices. A model with trend fits the data better, but the extra storage code needed to describe the trend parameter does not weigh against the improvement in the fit of the model. Rissanen (p. 97) concludes, wrongly, that this proves that 'any successful stock advice must either utilize inside information, which is not in the data, or it is entirely the result of luck'. While the quoted remark may be true, this cannot be inferred from the fact that the trend in the data has only limited descriptive power. A money-maker might still prefer the model with trend (but even then, transaction costs are probably higher than the expected returns).

3.4 Bounded rationality

The theory of bounded rationality, due to Simon (see Simon, 1986, for references), is currently experiencing a revival in game theory, as well as in the theory of learning in rational expectations models. Bounded rationality accepts the fact that people do not always seem to behave exactly according to the axioms of consumer behaviour and decision making under uncertainty. Situations of boundedly rational choice emerge for basically the same reasons as those that prevent use of the Solomonoff–Levin distribution directly. Some problems are not computable. In micro-economic phenomena, this bears on situations of strategic behaviour (recall Keynes' discussion of the beauty contest). In complex situations, one may opt for simple strategies. In macro-economics, disputes on the validity of some economic theories originate from different views on where exactly to locate bounded rationality: for example, consumers who suffer from money illusion, or entrepreneurs who are unable to discriminate between relative and absolute price changes. These implications for economic theory extend beyond the scope of this book, but there is a relation with (econometric) inference.

If the problem of deriving explicit and non-arbitrary prior probability distributions did not exist, then most philosophers (perhaps even statisticians) would agree that Bayesian inference is optimal: it results in coherent probability judgements and satisfies the likelihood principle. Bayesian prediction is an instance of Solomonoff's predictor which is considered ideal for the purpose of predictive inference (Li and Vitányi, 1992). However, these optimal qualities may be too good to be true. The Bayesian inference mechanism (conditioning, marginalizing and mapping) is akin to Laplace's demon.[13] The demon is empirically omniscient and knows the deterministic laws of physics (such as gravitation). A Bayesian demon, to paraphrase the metaphor, is logically omniscient by having an exhaustive set of hypotheses with their ideal universal prior probabilities. The Bayesian demon knows in advance how to respond to every new piece of information: there can be no surprises (this is similar to the complete markets hypothesis of an Arrow–Debreu economy, where it is possible to make perfect contingency plans).

Kolmogorov complexity theory suggests why this fiction of the Bayesian demon must be defective. Kemeny (1953, p. 711) expresses his doubts as follows:

> Few, if any modern philosophers still expect fool-proof rules for making inductive inferences. Indeed, with the help of such rules we could acquire infallible knowledge of the future, contrary to all our empiricist beliefs.

Laplace's demon did not survive the quantum revolution in physics. Similarly, the Bayesian demon is set back by Gödel's incompleteness theorem[14] or, which amounts to the same thing, the non-computability of the universal prior (Li and Vitányi, 1990, pp. 208–14, establish the link between complexity and Gödel's theorem). Even the most extreme 'objectivist' or 'necessarist' version of Bayesianism, based on the notion of a universal prior, must allow for a pinch of subjectivism.

The possibility of indeterminate decision problems, due to the non-computability problem, has been discussed by game theorists (e.g. Binmore, 1991). The problem is relevant for situations of strategic decision making. Examples are tax wars, interest rate and exchange rate policy on the macro level, or stock market speculation at the micro level. Complexity considerations affect the choice of strategies. Although outcomes of decisions need not be completely chaotic (namely random without following a probability law), they may well be.

4 Concluding comments

In this chapter, the concept of simplicity has been evaluated. Chapter 7 deals with its relation to some econometric methodologies, such as statistical reductionism. Reductionism holds that proper inference starts with a general model, or ideally, even the Data Generation Process (see chapter 7, section 4.2), and descends step by step to the more specific and simple model, until a critical line on loss of information is crossed. However, general to simple is an inversion of the history of scientific inference. Scientific progress in physics illustrates a sequence of increasing complexity: consider planetary motion, from circles to ellipses, to even more complicated motions based on relativity; from molecules to atoms, and to even smaller particles. Keplers' laws were mis-specifications, but were they invalid? The same question can be raised for cases in empirical econometrics, such as the investigation of consumer demand. The first econometric studies were based on very simple models, with few observations. Current investigations are not infrequently based on more than 10,000 observations, and are much more general in being less aggregated with more flexible functional forms. Were the pioneers wrong and misguided by their simple mis-specifications? Although reductionism does not pretend to be a theory of scientific development, such questions point to the defect it has as a modelling strategy. All models are wrong: Kepler's, Tinbergen's, Friedman's, and 'even' those of contemporary econometricians. Some, however, are useful. Kepler's clearly were, as were Tinbergen's and Friedman's. They shared a common respect for simplicity.

Notes

1. This chapter draws heavily on Keuzenkamp and McAleer (1995; 1997).
2. Thorburn (1918) argues that it is a modern myth. Although the precise formula as presented has not been found in Ockham's writings (but even today not all of his writings have been published), similar expressions can be found in the published writings of Ockham, such as '*pluralitas non est ponenda sine necessitate*' (plurality is not to be posited without necessity, see Boehner, 1989, pp. xx–xxi). The underlying idea concerning simplicity dates back to

Aristotle's *Physics*. Derkse (1992, p. 89) provides extensive background material and the intellectual history of Ockham's razor.

3. For example, Quine (1987, p. 12) gives '*entia non multiplicanda sunt praeter necessitatem*' (entities must not be multiplied without necessity). No explicit reference is given. Klant (1979, p. 49) presents 'Occam's Razor' as '*entia explicantia non sunt amplificanda praeter necessitatem*'. Rissanen (1983, p. 421) represents 'Ockham's razor' as 'plurality is not to be assumed without necessity'. Again, references are missing.
4. Thorburn (1918, p. 349) cites the third edition of the *Principia Mathematica* (1726, p. 387), where the formula '*natura nihil agit frustra, et frustra fit per plura quod fieri potest per pauciora*' appears, with no mention of Ockham. (Nature creates nothing for nothing, and nothing occurs with more if it could have occurred with less.)
5. Leibniz used the formula '*entia non esse multiplicanda praeter necessitatem*' in his inaugural dissertation (see Thorburn, 1918, p. 346), without specific reference to Ockham.
6. As an example, Jeffreys ([1939] 1961, p. 46) considers the hypothesis: $s = a + ut + \frac{1}{2}gt^2 + a_3t^3 + \ldots + a_nt^n$. If the first three terms on the right hand side can account for most variation in the data, it would be 'preposterous' to increase the number of adjustable parameters to or even beyond the number of observations: 'including too many terms will *lose* accuracy in prediction instead of gaining it'. Jeffreys (pp. 47–9) provides four examples for evaluating the complexity for differential equations, two of them derived from special cases of the equation above. Complexity is given by the sum of the order, degree and the absolute values of the coefficients. For example, for $s = a$, $ds/dt = 0$, which results in a complexity of $1 + 1 + 1 = 3$. For the case $s = a + ut + \frac{1}{2}gt^2$, he obtains $d^2s/dt^2 = 0$ (*sic* – according to Arnold Zellner, this is probably not a typo but due to the fact that Jeffreys considered the homogenous differential equation in this instance), which yields a complexity of $2 + 1 + 1 = 4$. A problem with interpreting these calculations is that Jeffreys does not define clearly what is meant by order, degree and the absolute value of the coefficients.
7. During a seminar presentation at Cambridge (UK), Willem Buiter complained that this is an extreme assumption. However, it is made in many rival approaches to inference. Wald (1940) relies on a similar assumption to repair von Mises' frequency theory. Wrinch and Jeffreys (1921) make it in the context of (necessarist) Bayesian induction. Popper ([1935] 1968) invokes it for his methodological falsificationism. Good (1968, p. 125), who is a subjective Bayesian, argues that parameter values can, in practice, take only computable values in the sense of Turing, provided that they are specified in a finite number of words and symbols. Ramsey (1974, p. 27; note, this is *not* Frank), who gives a frequentist approach to model discrimination, takes the number of models as countable for convenience.
8. A Turing Machine, named after Alan Turing, is basically a computer with infinitely expandable memory. A UTM is a Turing Machine which can simu-

late any other Turing Machine. If a program code (a finite string of 0s and 1s) c which is fed to the UTM yields S as output, then this is represented by $\text{UTM}(c) = S$. There may be many such strings c, but it may be the case that the UTM is unable to produce output. This is known as the halting problem.

9. Jeffreys' distribution has some undesirable properties, most importantly, it is 'improper' (it does not sum to 1). An additional advantage of the Solomonoff–Levin distribution is of course that in this case, a formally satisfactory definition of complexity is given. The formalization has been advanced independently by authors other than Solomonoff as well, in particular, by Kolmogorov – whence the current terminology of 'Kolmogorov-complexity'. See Li and Vitányi (1990) for a useful survey.
10. Rissanen (1983) shows that functions other than the $\log^*$ function also have the desired properties, but $\log^*$ is the most efficient to represent the integers. These functions share the property that the first term ($\log x$) is dominant.
11. Herbert Simon (like Solomonoff a student of Carnap while in Chicago) also worked towards the goal of automating science. In 1956, he and Allen Newell invented a computer program that could prove theorems of symbolic logic taken from Whitehead and Russell's *Principia Mathematica*, the Logic Theorist. Good (1965, p. 4) also argues that the goal of science is the automation of inference. A fine example of machine learning, based on using artificial intelligence techniques, can be found in Langley, *et al.* (1987). They discuss a number of computer programs based on contemporary information-processing psychology (cognitive science). The programs are able to derive, among others, Kepler's laws, using non-substantive heuristic rules (i.e. these rules do not impose theoretical restrictions that bear a direct relationship to the subject of study). Hence, it is a data-driven approach for discovering empirical laws.
12. Recall that the information criteria are derived by starting from prior ignorance about rival models. Once an investigator has been able to extract general information from specific data, the assumption of prior ignorance is no longer valid and informative probability distributions may be used.
13. Laplace's deterministic view on the character of the universe can be found in his *Théorie Analytique des Probabilités*, not in his *Traité de Mécanique Céleste*; see Suppes (1970, p. 32). Laplace uses probability theory for problems where ignorance and complex causes prevail.
14. Gödel's first incompleteness theorem states that there exist non-decidable problems (propositions which cannot be proved either true or false) in systems of formal logic.

6 From probability to econometrics

> It is a curious fact, which I shall endeavour to explain, that in this case a false hypothesis, which is undoubtedly a very convenient one to work upon, yields true results.
>
> Francis Galton (cited in Stigler, 1986, p. 274)

1 Introduction

What can we infer about a universe from a given sample? This is the recurring question of probabilistic inference. Bowley ([1901] 1937, p. 409), answers that the only bridge is the Bayesian one. In a footnote, he makes the distinction between the problem 'given the target how will shots be dispersed?' and 'given the shot-marks, what was the target?' The first question can be answered by means of Fisher's analysis of variance, the second 'is the practical question in many cases when we have only one sample' and necessarily depends on Bayesian inference (p. 409).

But the Bayesian argument never gained a strong foothold in econometrics: the frequency theory prevailed. The link between probability theory and practical statistics is sampling theory, which should provide the justification for using (parametric) regression techniques of inference.[1] But econometrics uses sampling theory more in a rhetorical than in the real sense. This chapter discusses the limitations of sampling theory in econometrics.

This chapter deals with how the different elements of probability filtered into econometrics. Section 2 briefly reconsiders the key elements of sampling theory. Section 3 is about regression. The contributions of Galton and Yule are discussed, including Galton's fallacy and 'nonsense correlation'. Section 4 is about the efforts of the early econometricians to justify the use of sampling theory.[2] Section 5 provides a summary of the advent of probabilistic inference in economics.

2 Sampling theory

Sampling theory provides a link between (frequentist) probability theory and statistics. Note for the additional step of epistemological inference, one needs *a priori* probabilities or an alternative bridge, like von Mises' second law of large numbers. Sampling theory is a major ingredient of 'textbook econometrics'. Consider the classic textbook of Goldberger (1964, p. 4). Econometric models, he claims, consist of three basic components:

- a specification of the process by which certain observed 'independent variables' are generated
- a specification of the process by which unobserved 'disturbances' are generated
- a specification of the relationship connecting these to observed 'dependent variables'.

The three components 'provide a model for the generation of economic observations, and implicitly, at least, they define a statistical population. A given body of economic observations may then be viewed as a sample from the population.' Goldberger (p. 67) interprets economic data as

> an observation on a random vector, viewing the datum as the outcome of some experiment; also [he] take[s] the empirical counterpart of the probability of an event to be the relative frequency of its occurrence in a long series of repeated trials.

Once the statistical population is defined, methods of probabilistic inference can be applied to the investigation of the economic relationships. Ordinary Least Squares (OLS) is a useful tool for this purpose, and it is discussed by Goldberger (pp. 161–2) in the context of the classical linear regression model. In his case, this consists of OLS plus the assumptions of linearity, zero expected value of the 'disturbance', constant variance and uncorrelated residuals. Finally, he (p. 162 and elsewhere) takes the regressors as 'fixed in repeated samples'. The notion of repeated sampling enters the motivating discussion which precedes the formal presentation of regression.

A conditional distribution is introduced as the main feature of the classical linear regression model. The observations are regarded as drawings from the appropriate conditional distribution. Goldberger (p. 161) continues:

> A convenient alternative formulation is: There are T populations of random disturbances; each has mean zero, all have the same variance. Each joint observation is a given set of values of the x's together with a value of y determined as a constant linear function of the x's plus a drawing from a disturbance population. Successive drawings of the disturbance are uncorrelated. The T sets of regressor values are fixed in repeated samples.

Of course, these are all assumptions. No further justification for the validity of those assumptions in economics are given. The applicability of sampling theory was taken for granted in this 'textbook' era of econometrics.

Before discussing sampling and regression in econometrics (which will be done in the subsequent sections), the basic elements of sampling theory need to be discussed in more detail (see for more details Mood, Graybill and Boes, 1974, pp. 222–36). To start with, a *target population* is defined as the population of elements about which information is desired. Pudney (1989, p. 48) calls this the 'super-population' (the counterpart to von Mises' collective or Fisher's hypothetical infinite population). Consider the definition of a random sample. Let x_i $(i = 1, \ldots, n)$ be a realization of some random variable. Then a random sample is a set of n observations $c\,x_1, \ldots x_n)$ if the joint density factors as:

$$f(x_1, \ldots x_n) = \prod_{i=1}^{n} f_i(x_i). \tag{1}$$

This reflects the independence of drawings. In addition, unbiasedness is occasionally required (e.g. Pudney, 1989, p. 48), implying that each observation has the same distribution as the (super-) population. Mood, Graybill and Boes, 1974 note that, in practice, the target population may be hard to sample. In that case, a *sampled population* is invoked which should be informative about the target population. Whether it is, will often be a matter of judgement. For example, children of 'bad stock' in Manchester may not be representative for children in Britain, as Keynes ([1910] 1983, CW XI, pp. 186–216) argued in an early dispute with Karl Pearson.[3]

Two justifications are sometimes (wrongly) given to apply sampling theory: the law of large numbers and the central limit theorem. Recall that the law of large numbers means that, for an arbitrary small difference between the 'true' population mean and the sample mean, the probability that the difference is indeed that small can be made as close to one as desired by increasing the sample size. A condition for the (weak as well as strong) law of large numbers is that a random sample is at stake. This has to be justified. Even so, the law does not provide a 'bridge' from sample to population even given a truly random sample.

The central limit theorem (due to Laplace) provides information about the distribution of a sample mean of a random sample. It states that, for a sample of independent random variables X_i $(i = 1, \ldots, n)$ each with mean μ and finite variance σ^2, the standardized sample mean has a distribution that approaches the normal distribution. Again, randomness is a necessary condition for the validity of the theorem and this cannot be taken for granted on *a priori* grounds. Simply invoking a normal distribution in empirical work by making reference to the central limit theorem begs the question of the justification of sampling theory. Still, the theorem is remarkable as no information is required about the original density function of the random variables.

The only real justifications for using sampling theory are either sample design, for example by means of experiments, or external evidence (prior information) on the validity of sampling assumptions. An (invalid) alternative, the design of the characteristics of residuals in regression equations by means of statistical modelling, will be discussed at length in the next chapter.

3 Regression

3.1 Quetelismus

Applications of probability theory to economics were rare until the beginning of the twentieth century. The main difficulty for applying statistical inference in the social sciences was the problem of classifying data in homogeneous groups (Stigler, 1987). Such groups are categories for which the major influential factors can be considered constant, and where residual variation is due to chance. Homogeneity is often problematic. The characteristics of deaths in rural areas may be unlike those in cities, for example. The problem for the social sciences is that there is no *a priori* or logical rule for the classification of homogeneous groups. This is also Keynes' complaint of econometrics: in many cases, there is no reason why one should regard a sample as homogeneous.

Adolphe Quetelet is regarded as a 'parent of modern statistical method' (Keynes, [1921] CW VIII, p. 367). He is known for fitting 'normal curves' (curves based on the normal probability distribution) to a wide range of data, such as the length of different groups of the population. According to this Belgian statistician, if a normal curve is observed, it must be due to a constant cause and random fluctuations of independent accidental causes – a reversal of Laplace's central limit theorem. Quetelet exaggerates the prevalence of the normal 'law'. He reports normal curves for all kinds of classifications, to such an extent that his efforts to find normal curves discredited their application to the social sciences. It became known as 'Quetelismus'. Quetelet did not know that a compounding of normal distributions results in another normal curve: this was discovered by Galton who illustrated his finding with the 'quincunx' or Galton's Board (see Stigler, 1986, pp. 276–80), a pinboard that creates normal distributions with small balls. Comte, the founder of positivism, concluded from Quetelet's efforts that applying statistics to the social sciences yields a chimera and should be rejected. Keynes ([1921] CW VIII, p. 367) complains that this parent of statistics 'prevented probability from becoming, in the scientific salon, perfectly respectable. There is still for scientists a smack of astrology, of alchemy.' The work of Galton, Yule and subsequently the fundamental innovations of Fisher and other probability theorists, changed this state of affairs.

3.2 Curve fitting: least squares, regression and a fallacy

The method of least squares was invented in 1805 by Legendre, who used it in order to fit orbits of comets to (linear) equations. Consider a vector of observations $\mathbf{y}$ ($nx1$), related to the matrix of observations of k variables $\mathbf{X}$ (nxk), where $k = 1$ in the early stage of statistics. Legendre's problem (in modern notation) is to obtain a system of equations $\mathbf{y} = \mathbf{X}\beta + \mathbf{u}$, 'to be determined by the condition that each value of $\mathbf{u}$ is reduced either to zero, or to a very small quantity' (Legendre, quoted in Stigler, 1986, p. 13). The problem becomes intricate if the number of individual equations (observations) is larger than the number of unknown parameters. Legendre proposes to decide the problem by the method of '*moindres quarrés*':

> it consists of making the sum of the squares of the errors a minimum. By this method, a kind of equilibrium is established among the errors which, since it prevents the extremes from dominating, is appropriate for revealing the state of the system which most nearly approaches truth.

Hence, Legendre suggests to minimize $\mathbf{u}'\mathbf{u}$, which yields the familiar least squares solution,

$$\mathbf{b} = (\mathbf{X}'\mathbf{X})^{-1}\mathbf{X}'\mathbf{y}. \tag{2}$$

Note that Legendre aims at approximating 'truth' by non-stochastic means: his problem is not related to probabilistic inference (Rudolf Kalman continues this honourable tradition today; his impact on econometrics is not large, however). No interpretation of probability is needed to apply the method of least squares. The terminology 'least squares estimation' is therefore misleading, although widely accepted.

Gauss ([1809] 1963) gives a probabilistic argument which, remarkably, yields the same result as Legendre's. Gauss studies a problem similar to Legendre's (i.e. orbits of planets), but argues that the errors to the set of linear equations must have a probability distribution. Among others, this distribution should be symmetric and have its maximum at $\mathbf{u} = 0$. From these conditions, Gauss postulates the normal ('Gaussian') distribution. Gauss shows that the estimators that follow coincide with the least squares approximation. In a later publication, Gauss provides the basic ingredients of what is now known as the Gauss–Markov theorem: in modern parlance, his method yields an estimator with minimum variance within the class of linear unbiased estimators (Stigler, 1986, p. 148).

With some delay, Gauss' method conquered economics.[4] The delay would have been longer without the accomplishments of Galton (simple regression analysis) and Yule (multiple regression analysis). Galton's innovation is the application of the normal distribution for studying the association between two random variables. The new insight relative to Quetelet is that Galton starts to understand how a mixture of different distributions might lead to one normal distribution: the underlying factors do not have to be homogeneous, but the resulting distribution obeys the 'law of errors' and allows statistical analysis (see Stigler, 1986, p. 275, where this is illustrated with Galton's 'quincunx': this mechanical pinboard transforms a mixture of binomial distributions into a normal one). This sets the basis of a model of conditional inference in the social sciences.

In one of his most famous applications, Galton finds a positive relation between heights of children and their parents, plus a 'regression to mediocrity' (movement to the average; see Stigler, 1986, pp. 294–5; Yule, [1911] 1929, pp. 175–7). However, this 'regression' is an artifact of the method Galton used, which yields a bias towards zero. It is known as Galton's 'regression fallacy', first recognized by Karl Pearson.

Consider figure 2, which depicts a scatter of points $(\mathbf{x}, \mathbf{y})$ (which in Galton's case would represent parents' stature on the horizontal axis and children's stature on the vertical axis). The regression model assumes that there is a relation between $\mathbf{x}$ and $\mathbf{y}$ for which equation (2) is the least squares solution (this is the regression line in the scatter, which is flatter than the major axis through the concentration ellipse). This solution is based on the assumption that $\mathbf{y}$ is noisy (or random), whereas $\mathbf{x}$ is noise free. Alternatively, it may be assumed that $\mathbf{x}$ rather than $\mathbf{y}$ is noisy, or that both are. This boils down to the question whether $\mathbf{y}$ should be regressed on $\mathbf{x}$ or conversely. These regressions yield different estimates for β. Galton (who recognized but then ignored the problem) assumed that the children's stature was normally distributed, whereas parents' stature was not stochastic.

As a result, he concluded that children's stature 'regressed' to the mean: $\beta < 1$. If his conclusion were true, then during history the variance in human height should have been reduced over time, *quad non*. The conclusion is based on a fallacy due to confusing a (valid) conditional

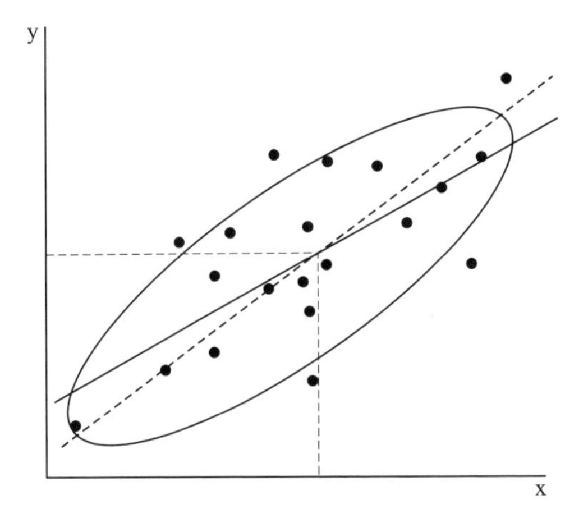

Figure 2

expectation with a general tendency. This confusion results from the stochastic assumptions involved. In order to investigate a general tendency, one should take care not only of the measurement error in **y**, but also on the measurement error in **x**. Assume both **x** and **y** are measured with means μ_x and μ_y respectively, and variances σ_x^2 and σ_y^2. Their correlation is denoted by ρ. Then the conditional distribution of **y** given **x** is normal with mean

$$E(\mathbf{y}|\mathbf{x}) = \mu_y + \rho\frac{\sigma_y}{\sigma_x}(\mathbf{x} - \mu_\mathbf{x}). \tag{3}$$

In case of the stature example, one may assume for sake of simplicity that σ_x and σ_y are equal, in which case (3) shows that the conditional expectation of **y** depends on the amount of deviation of observervations of **x** from their (unconditional) mean, but most likely less than proportional since $|\rho| \leq 1$ and the strict inequality may be expected to hold. Note that exactly the same argument can be made for the inverse regression.

For Ragnar Frisch, who made an effort to analyse this problem (Frisch, 1934), the problem was an important reason to reject probabilistic econometrics (it is not a well-known fact that one of the founding fathers of econometrics and co-winner of the first Nobel prize in economics was highly critical of econometrics). His alternative to ordinary regression, 'bunch map analysis', is a somewhat cumbersome way of investigating the sensitivity of the estimates to changing the allocation of stochastics to different variables. Koopmans (1937) provided an analytical solution to the 'errors in variables' problem (as he called it). However, Koopmans' solution hardly received the attention it should have had – an important reason is that it was overshadowed by the 'errors in equations' problem that was put on the agenda by Haavelmo and adopted as the core of the Cowles programme in econometrics – including Koopmans.

Although the regression fallacy has been long recognized, 'old fallacies never die' (Friedman, 1992). Research on convergence between economies in the recent empirical growth literature suffers from the same old problem as Galton's inferences (see Quah, 1993). Frisch and, more recently, Kalman criticize econometrics for the neglect of this fallacy, and argue that the probabilistic approach is invalid. A less extreme conclusion is that econometricians ought to consider their assumptions more carefully.

After Galton, the next important innovation which made Gauss' method useful in the social sciences is multiple regression. This is Yule's contribution. He is also the first one to apply it to economics. In his study on pauperism, Yule (1899, cited in Stigler, 1986, pp. 355–6) argues that the regression coefficient of some parameter of interest in a multiple regression equation 'gives the change due to this factor *when all the others are kept constant*'. Multiple correlation is the way to deal with the *ceteris paribus* clause (section 3.4, below). Yule's study provoked a critique by Arthur Cecil Pigou, who testified in a 1910 memorandum to a Royal Commission on poor law reform. This memorandum contains a section on 'The Limitations of Statistical Reasoning'. Pigou (quoted in Stigler, 1986, p. 356) complains that

> some of the most important influences at work are of a kind which cannot be measured quantitatively, and cannot, therefore, be brought within the jurisdiction of statistical machinery, however elaborate.

In fact, Pigou claims that Yule's results are without value, due to omitted variables: the *ceteris paribus* do not apply. This criticism may have been less than persuasive to Yule. Although Pigou points to an omitted variable (management quality of social security administrations), he does not substantiate his claim that this variable really does account for some of the results. In the exposition on correlation in Yule ([1911] 1929), where the pauperism example is used as an illustration, Pigou is not mentioned.

Yule's investigation of pauperism is nearly forgotten today, but his methods of inference are not. Stigler suggests that only after the introduction of multiple regression techniques, did the classification problem stop being a stumbling block for the applications of probability techniques for the 'measurement of uncertainty' in the social sciences. This is overly optimistic, though. When discussing 'Professor Tinbergen's method', Keynes ([1939] 1973, CW XIV, p. 316), for example, still objects to applying multiple correlation to the social sciences,

> the main *prima facie* objection to the application of the method of multiple correlation to complex economic problems lies in the apparent lack of any adequate degree of uniformity in the environment.

But it is undeniable that multiple regression techniques took off in economics.

3.3 Spurious and nonsense correlation

Another problem caused by the invalidity of sampling assumptions is 'spurious correlation'. The name is due to Karl Pearson (in an 1897 discussion of correlation between indices; see Yule, [1911] 1929, p. 226, for references). The most famous illustration comes from Neyman (1952, pp. 143–50), who presents the statistical 'discovery' that storks bring babies. A sample of raw data for fifty-four counties is given, with, for each county, the number w of fertile women, the number of storks, s, and the number of births in some period, b. As the number of births is likely to be higher in larger counties, where more storks as well as fertile women live, regressing b on s will not yield a correct inference. One should correct for county size. This can be done by creating the transformed variables $x = s/w$ and $y = b/w$. These variables are clearly correlated in Neyman's sample, although the correlation is not very plausible – spurious, Pearson would say. Neyman has proposed (rightly, though in vain) speaking of a 'spurious method' rather than 'spurious correlation', as the problem is the way the variables are adjusted for county size: the variables x and y depend on a common third variable, w. The correlation is real. The influence of this third factor should be eliminated before concluding that storks bring babies (Spanos, 1986, gives an elaborate discussion of the omitted variable argument).

A special case of spurious correlation is Yule's nonsense correlation (although Yule himself regards nonsense correlation as a completely distinct case). 'Nonsense correlations' is the standard terminology for high correlations between independent but integrated (highly autocorrelated) time series.[5] Yule (1926) is commonly cited as the first paper on nonsense regression, but he and other statisticians were aware of nonsense correlation long before (e.g. Yule's study on pauperism, which dates back to 1899, uses relative changes as its variables; Yule, 1921, discusses the issue in the context of differencing and detrending time series). Yule (1926) observes that there is a very high correlation between standardized mortality per 1,000 persons in England and Wales, and the proportion of Church of England marriages per 1,000 of all marriages, if the period 1866–1911 is investigated. He (p. 2) argues:

> Now I suppose it is possible, given a little ingenuity and goodwill, to rationalize nearly anything. And I can imagine some enthusiast arguing that the fall in the proportion of Church of England marriages is simply due to the Spread of Scientific Thinking since 1866, and the fall in mortality is also clearly to be ascribed to the Progress of Science; hence both variables are largely or mainly influenced by a common factor and consequently ought to be highly correlated.

But Yule shows that the high and significant correlation between the two variables is entirely due to the autocorrelation of the time series, an important new insight.

Unlike his tutor, Karl Pearson, Yule is interested in causal inference (see chapter 9, section 3, below). This is why he is unwilling to interpret the problem as one of spurious correlation where a third 'causal factor' is ignored, as it is not correct to 'regard time *per se* as a causal factor' (Yule, 1926, p. 4). The real problem, he argues, is that the time series that yield this nonsense correlation do not in the least resemble 'random series as required' (p. 5). It is autocorrelation, lack of randomness, that generates nonsense correlation.

Although serial correlation is much better understood today, how to deal with it remains controversial. The frequentist interpretation of probabilistic inference already runs into severe problems for the relatively simple integrated (I(1)) time series. Whether it is necessary to detrend, difference, or otherwise filter economic time-series is not obvious, despite the vast co-integration literature which emerged in the 1980s.

3.4 Multiple regression and ceteris paribus

A well known problem of scientific inference is the justification of the *ceteris paribus* ('other things being equal') clause, which is nearly always invoked or implicitly assumed. Multiple correlation has been regarded as a way to deal with the 'other things': control for them in such a way that inference with respect to the variable(s) of interest is valid.

An early critic of 'infant' econometrics was Alfred Marshall. In a letter, dated 16 January 1912 (cited in Mirowski, 1989, p. 222), he criticizes research of the pioneer of consumer demand econometrics (see chapter 8, below), Henry Ludwell Moore:

it proceeds on lines which I had deliberately decided not to follow years ago; even before mathematics had ceased to be a familiar language to me. My reasons for it are mainly two.
(1) No important economic chain of events seems likely to be associated with any one cause so predominantly that a study of the concomitant variation of the two elements with a large number of other causes representing the other operative causes: the 'caeteris paribus' clause – though formally adequate seems to me impracticable.
(2) Nearly a half of the whole operative economic causes have refused as yet to be established statistically.

Moore (a student of Karl Pearson) does not agree. Applying modern statistical methods, 'the calculus of mass phenomena', is feasible because of the emergence of abundant data, he argues. Of course, in reality Moore and the other early econometricians had very limited data sets (and computer facilities), certainly not of the size requested, for example, by von Mises.

Marshall's letter may have provoked an explicit treatment of *ceteris paribus* in Moore (1914, p. 67). Here, he regards the method of multiple correlation as a good alternative to the 'unrealistic' *ceteris paribus* clause, as it only assumes that 'other things' change according to their natural order. Still, Marshall's argument has been repeated by many critics of econometrics, including Pigou, Keynes and Robbins. Proper controlling for other influences is possible only if all confounding influences are known, and if the relevant data are available. Neither economic theory nor measurements are sufficiently strong to make the assumption credible that everything that is relevant (all nuisance variables) can be controlled for. And economic theory rarely provides a justification for a specific probability distribution of the residuals in econometric equations. In most cases, the error term in econometric models is added as an afterthought.

Hence, economic theory should be assisted by experimental design. Randomization in experiments is an alternative way to deal with unknown elements in the *ceteris paribus* clause (see chapter 3, section 3.3). For Fisher, randomization takes care of the *ceteris paribus* clause. A statistical model does not contain more than a few of the most important factors that interest a researcher. Many other causes may be effective and bias the experiment to an unwarranted conclusion. If the experiment is properly randomized, Fisher ([1935] 1966, p. 44) writes, this may relieve

the experimenter from the anxiety of considering and estimating the magnitude of the innumerable causes by which his data may be disturbed. The one flaw in Darwin's procedure was the absence of randomization.

Fisher's theory goes beyond Yule's multivariate regression. Whereas Yule thought that the *ceteris paribus* condition could be satisfied by controlling for the most important influences, Fisher recognizes that this alone will not do to avoid wrong inferences.

Experimental design is indeed a 'buzz-word' in the classic foundational paper of modern econometrics (Haavelmo (1944)). His use of experimental design will be discussed in the next section, here the focus is on the *ceteris paribus* clause. In studying what actually happens, Haavelmo (pp. 16–17) argues, the art is to construct models 'for which we hope that Nature itself will take care of the necessary *ceteris paribus* conditions'. This hope is based on knowledge of the past – induction, based upon a synthetic *a priori* belief in the stability of nature. He (p. 17) also discusses what would happen if other things do *not* remain equal. In that case,

> we try to take care of the *ceteris paribus* conditions ourselves, by statistical devices of clearing the data from influences not taken account of in the theory (e.g. by multiple-correlation analysis).

Does this mean that the econometrician has to include variables which do not follow from (economic) theory in the model, in order to satisfy the assumptions for valid statistical inference? If this is the case, then non-theoretical considerations are crucial in justifying the inductive aims of the econometrician, which is remarkable in light of the strong reliance of Haavelmo and his followers on economic theory for specification purposes.

4 Samples and populations in econometrics

Random sampling is meant to provide the justification for inference in the frequentist (sampling) theory of probability, given prior knowledge of the underlying probability distribution. This prior knowledge may derive from theoretical knowledge, or from the set-up of some experiment. If this distribution is not known, and moreover, controlled experimentation is impossible, inference becomes problematic. Regression, the much used technique in econometrics, suffers from these problems. Econometricians have tried to provide a sampling theoretical framework by invoking economic interpretations of the notions 'sample' and 'population'. Some notable contributors to the literature are discussed in this section.[6] They can be divided into three groups:

- *a priori* synthetic
- analytical
- metaphorical justifications.

The first approach invokes a principle of uniformity of nature. The second is based on extending statistical theorems, in particular the central limit theorem or the law of large numbers. The third approach invokes analogues which are thought to support the use of probabilistic inference in economics.

4.1 A priori *arguments*

First consider an *a priori* justification. In a chapter on 'Statistical induction and the problem of sampling', Mills (1924, p. 550) poses the following question:

> A statistical measure – an average, a frequency ratio, a coefficient of correlation has been derived from the study of certain data drawn from a large population. (The term 'population' refers to a group of things or phenomena having, presumably, certain characteristics in common.) May we assume that, if additional samples were taken from the same population, the corresponding measures would have the same values?

For von Mises, such a question about the stability of statistical results should be settled by empirical experience. Mills (pp. 550–1; also 1938, pp. 454–7), who refers to von Mises elsewhere (e.g. Mills, 1938, p. 732), takes a different approach:

> The statistical measures secured from successive samples might be assumed to be stable if the validity of two prior assumptions be granted. The first of these is, in general terms, an assumption of uniformity in nature, the assumption that there is, in nature, a 'limitation to the amount of independent variety'.

This is one of the rare references to Keynes' principle of limited independent variation in the econometric literature. But Keynes never took this principle for granted, and neither would von Mises. Mills' (1924, p. 552) second assumption is that the sample must be 'thoroughly representative of the entire population to which the results are to be applied'. Subsequently, Yule's assumptions of simple sampling are listed, of which independence is the most important one.[7] However, no further justification is given for accepting these assumptions. Mills (1938, p. 487) only acknowledges that, when dealing with time series, we are 'drawing from different universes'. Econometricians trying to grab the meaning of representative samples were not much helped by Mills.

4.2 Analytical arguments

The textbook of Crum, Patton and Tebbutt (1938) invokes a more analytical argument. Crum *et al.* (p. 216) present a verbal formulation of the definition of a random sample presented above, see (1). They continue by discussing sample design:

> In practice such sample is usually chosen by chance; but, as the manner in which the element of chance may vary, not all selections by chance are truly random samples. In contrast to the random sample is the *sample by design*, in which the selection is made upon some arbitrary plan designed to yield a sample which is representative of the population. Although sampling by design is used widely in practical statistics, it is only to *random* sampling that the doctrine of chance – the theory of error – can properly be applied.

This may pertain to a misunderstanding of Fisher, who argues correctly that experimental design or sample design yields the justification for statistical inference.[8] Alternatively, either the authors are unaware of Fisher's theory of randomization, or they disagree with Fisher's theory (which indeed was not generally accepted at the time they were writing). The latter possibility cannot be ruled out, but no explicit justification for disagreement is provided. However, Crum, Patton and Tebbutt continue with a remark that holds water:

> Another term sometimes applied to a sample is *representative*; but no generally accepted precise definition is available, and it is often used as a 'weasel word' by those who have no critical knowledge of the sampling method used.

They rightly claim that 'representative' was not well understood at the time. One may wonder whether it is today. Only recently, econometricians have started to analyse, for example, the problems resulting from selection bias in cross-section data. Crum, Patton and Tebbutt recognize the caveats of sampling in econometrics. Their justification for applying sampling theory, based on a reference to the central limit theorem (p. 220), is analytical but fallacious.[9]

Sampling theory remained enigmatic to economic statisticians. Ezekiel (1930, p. 13, n.1) writes that the terms 'universe' and 'sample' should be clearly understood. But in the first (1930) edition of his textbook Ezekiel's understanding was limited (though not always wrong), while in the second (1941) edition (which deals more explicitly with time series) Ezekiel is aware of some caveats but also contributes to the confusion. Ezekiel is primarily known for his research in empirical agricultural economics and for devising appropriate research methods. The kinds of problems investigated by him were similar in kind to the problems that occupied Fisher at Rothamsted Experimental Station. Examples given in Ezekiel (1930) deal with the relation between feeding methods of cows and output of dairy farms, farm size and income, or rainfall and crop yield. Most problems have a cross-section character. These problems in agricultural economics may be analysed by means of Fisher's methods, especially if his views on experimental design are taken into account.[10] Agricultural economics is a case where the notions of sampling, and possibly a stable universe, may make sense. But Ezekiel (e.g. p. 219) emphasizes the intricate nature of the underlying assumptions of his statistical investigations, in particular the assumption of homogeneous samples.

Ezekiel (e.g. p. 236) treats the few examples with a time-series character the same as the cross-section cases. The question, 'which population is at stake?' is not discussed. This changes in Ezekiel (1941) which contains an added chapter on 'The reliability of an individual forecast and of time-series analyses'. Here (p. 341), he acknowledges that in cases 'involving successive observations in time, theories of simple sampling do not apply rigorously, since the observations are not drawn fully at random'.[11] Therefore, he discusses problems caused by serial correlation, and problems due to the 'lack of constancy of the universe'. He continues,

> The formulas of simple sampling assume that there is a large or infinite universe of similar events, from which the sample is drawn at random. Such a universe might be, for example, the number of dots turned up at each throw by throwing a pair of dice a large number of times. They also assume that new observations or new samples will be obtained by drawing in exactly the same way from exactly the same universe, as by making additional throws with the same set of dice under exactly the same conditions... When any phenomenon is sampled at successive intervals of time the 'universe' being studied can never be precisely the same. (p. 350)

The problem of serial correlation is an additional complication. It cannot be just solved by allowing for trend factors. Indeed Ezekiel (p. 352) concludes,

> This is one of the greatest unsolved questions in the whole field of modern statistical methods. It is one where the widest possible range of judgements may be found among the experts who should be able to agree upon the answer.

Ezekiel is unable to present convincing analytic arguments for the validity of statistical inference in such occasions, but provides examples of successful applications (e.g. fairly accurate extrapolations that can be made in time-series analysis). The discussion by Ezekiel is concluded by a few remarks on the applicability of correlation methods in cases of changing universes. All being said, it remains, in his view, the 'baffling problem of changing conditions in time series' (pp. 426–7).

4.3 Metaphorical arguments

Modern econometrics is based on the more rigorous work of Koopmans, Haavelmo and some other econometricians who worked on analytical foundations of econometrics during the late 1930s and the 1940s. Have they been able to solve the 'baffling' problems which bothered the pioneers of econometrics?

Consider Koopmans (1937). He takes Fisher's theory as his starting point. Koopmans is much more explicit on the justification for using probabilistic methods in economics than his teacher, Tinbergen.[12] It is fair to say that Tinbergen remains close to the 'least squares approximation' interpretation, while Koopmans really tries to take the stochastics (i.e. estimation and regression) seriously. But Koopmans (p. 5) is aware of the fact that

> In economic analysis variables at the control of an experimenting institution are exceptional... At any rate, they are far from being random drawings from any distribution whatever.

Despite this acknowledgement, he (p. 2) expresses his debt to Fisher's method. He notes that it not only relies on the notion of a hypothetical infinite population, but also on the possibility of repeated sampling (in fact, he should have stressed experimental design instead of repeated sampling, as Fisher did not regard repeated sampling as a necessary condition for inference).[13] However, Koopmans (p. 7) in turn qualifies its importance:

> the expression '*repeated sampling*' requires an interpretation somewhat different from that prevailing in applications of sampling theory in the agricultural and biological field.

Koopmans notes that, in those fields, the distribution of statistics can be obtained from experimentation, as repeated samples may be obtained in any number. Fisher's (1922, p. 311) 'hypothetical infinite population, of which the actual data are regarded as constituting a random sample' is, therefore, not quite comparable with the imaginary population that underlies economic data. Koopmans (1937, p. 7) writes,

> in the conditions under which sampling theory is used here such a distribution is much more hypothetical in nature. The observations [$X_k^{(t)}$, $k = 1, 2 \ldots K$, $t = 1, 2, \ldots T$] constituting one sample, a repeated sample consists of a set of values which the variables would have assumed if in these years the systematic components had been the same and the erratic components had been other independent random drawings from the distribution they are supposed to have.

Koopmans (p. 64) considers the sample as one single observation, and only one observation can be obtained for a particular time period. Because of the scarcity of economic data, Fisher's small sample theory is supported, and the maximum likelihood method applied.

Koopmans (pp. 26–7) adopts the standard interpretation of significance testing, where also the confidence interval for some parameter β_k, is explained:

> it cannot be said that this statement has, in the individual case considered, a probability $1 - \theta$ to be true. For a definite probability cannot be attached, in the individual case, to any statement of this kind concerning β_k unless an a priori probability distribution of β_k is known. The relevant fact about the above statement is, however, that, if it is repeated again and again in a large number of cases in each of which the specification of this section applies, there will be a risk of error θ inherent in that procedure.

Koopmans provides the correct interpretation, but does not say when it holds in the domain of interest: the analysis of economic data. Instead, he just refers for further considerations of this issue to Fisher's writings on inverse probability and inductive inference as well as a publication of Neyman on this theme.

An argument to justify a probability approach in econometrics similar to that of Koopmans is used by Haavelmo (1944). He goes a step further, by claiming that some kind of repetition is conceivable. Haavelmo's argument merits extensive citation:

> There is no logical difficulty involved in considering the 'whole population as a sample', for the class of populations we are dealing with does *not* consist of an infinity of different individuals, it consists of an infinity of possible *decisions* which might be taken with respect to the value of *y*. And all the decisions taken by all the individuals who were present during one year, say, may be considered as one sample, all the decisions taken by, perhaps, the *same* individuals during another year may be considered as *another* sample, and so forth. From this point of view we may consider the total number of possible observations (the total number of decisions to consume *A* by all individuals) as the result of a sampling procedure, which *Nature* is carrying out, and which we merely watch as passive observers.[14] (p. 51)

The justification of treating unique economic data as random drawings resulting from an imaginary sampling procedure, is paralleled in the social sciences in the same period. Stouffer (1934, p. 481) may even be the first to have used the notion of 'imaginary samples' out of a 'population, or universe, of possibilities'. Such a universe, he writes, is a conception which 'is taking the central place in modern physics, although its validity is not undisputed'. Like Haavelmo, Stouffer applies his metaphor to the time series domain. His example is interesting: where Yule might have used it ironically, Stouffer seems serious when he says that it

> is applicable equally logically to a time series revealing, for example, certain synchronous fluctuations over a period of fifty years in England and Wales between indexes of business conditions and the marriage rate. The fifty consecutive pairs of entries represent a sample of the vast number of entires which might have been thrown up in this fifty-year period.

He continues (p. 482):

> Some timid temperaments among us may prefer to dismiss the conception, calling it moonshine. But is it moonshine?... Who would vouchsafe a categorical answer with much confidence at the present time?

Haavelmo, perhaps?

Note the subtle difference between Fisher's possible trials and the notion as it is used by Stouffer and Haavelmo. In their case, the trials are at best imaginary but certainly not possible in the common English understanding of the word. British empiricism, to which tradition Fisher belongs, is replaced by metaphysics, when Haavelmo writes that economic theory will provide the 'hypothetical probability models from which it is possible, by random drawings, to reproduce samples of the type given by Nature'.

Haavelmo's view is an interesting mixture of Fisherian sampling and experimental design theory and Neyman–Pearson testing (an additional influence is Wald). Haavelmo (1944, p. 8) uses the rethoric of experimental design, without referring to Fisher:[15]

> the model will have an economic meaning only when associated with a design of actual experiments that describes – and indicates how to measure – a system of 'true' variables (or objects) $x_1, x_2, \ldots x_n$ that are to be identified with the corresponding variables in the theory.

The phrase 'design of experiments' occurs nineteen times in Haavelmo (1944, pp. 6–18). The conclusion is that, if only economists would be careful in specifying the appropriate design of experiments that should accompany a particular theory, the confusion on the constancy or invariance of economic 'laws' would disappear (p. 15). Haavelmo's use of 'design of experiments' is, however, at odds with Fisher's. For Haavelmo, experimental design provides the bridge between theoretical terms and observational entities (like the correspondence rules of the logical positivists). This is along the lines of the Harvard philosopher Percy Bridgman, who introduced the term 'operationalism' for this purpose.

Fisher's contribution to Haavelmo's blend of probability theory is more apparent than real, as far as experimental design is concerned. Haavelmo's frequency interpretation of the theory of probability is less deceptive. He starts from the measure-theoretical formulation of probability, which in principle is interpretation-free (probability is a primitive notion). Haavelmo (p. 9) gives a frequentist interpretation to this primitive notion because it 'has been found fruitful in various fields of research' (p. 47). Like von Mises, 'experience' is the guide to appicability. Von Mises, however, is very strict in assessing when 'experience' warrants the frequency interpretation: small samples common in economics would not qualify. The experience of Keynes is that economic data are not 'homogeneous' and even if they were, the frequency approach is invalid because of its logical problems. Apparently, the appropriateness of economic experience for a frequency theory is either in the eyes of the beholder or a matter of logical dispute. Haavelmo does not elaborate on why 'experience' in economics warrants a frequency interpretation, except that, after asking whether the model will hold for future observations, he answers (p. 10):

> We cannot give any a priori reason for such a supposition. We can only say that, according to a vast record of actual experiences, it seems to have been fruitful to believe in the possibility of such empirical inductions.

This is not a minor exaggeration: probabilistic economic inference in economics was in its infancy. Worse, the kind of methods used in this 'prehistory' of econometrics were severely criticized as being close to meaningless: without probabilistic foundation, without dealing with simultaneity problems. How could they have proven the success of his approach?

Although Haavelmo at first supports the frequency interpretation, he is not perfectly consistent. At one point, he gives modest support to what may be regarded as an implicit Bayesian interpretation of statistics, not unlike Fisher's interpretation of probability as a measure of evidence:

> we considered 'frequency of occurrence' as a practical counterpart to probability. But in many cases such an interpretation would seem rather artificial, e.g., for economic time series where a repetition of the 'experiment', in the usual sense, is not possible or feasible. Here we might then, alternatively, interpret 'probability' simply as a measure of our *a priori confidence* in the occurrence of a certain event. (p. 48)

But the econometrician should not make explicit use of inverse probability, as hypotheses are either true or false, in which case a probability distribution for the hypotheses space 'does not have much sense' (p. 62).

This leads to Haavelmo's step of introducing the Neyman–Pearson methodology in econometrics. Explicitly, this time (see p. iv). However, Haavelmo continues to make use of the standard notion 'inference' instead of using the behaviouristic language of Neyman–Pearson.

In view of these observations, it is perhaps no surprise that Haavelmo dismisses the discussions on the interpretation of probability by saying that this has resulted in 'much futile debate' (p. 48). He concludes (p. 48):

> The rigorous notions of probabilities and probability distributions 'exist' only in our rational mind, serving us only as a tool for deriving practical statements of the type described above.

Hence, this effort to create foundations for the probability approach in econometrics finally results in an inconsistent set of claims in its defence. First, there are vast amounts of experience which warrant a frequency interpretation. This is supported by repetitive discussions of experimental design, but the inability to experiment inspires an epistemological interpretation. Then Haavelmo mentions the futility of bothering with these issues because the probability approach is most of all a useful tool. This would be an instrumentalistic justification for its use if Haavelmo gave supportive evidence for his claim. There is not one example which attempts to do so.

5 Conclusion

Is all the talk about experiments in econometrics just noise? No. Two recent developments may bring econometrics a valid justification of the Fisher-paradigm. They are 'real' experiments, and the 'natural experiment'. The first becomes increasingly popular in game theory, while the second has been proposed for causal analysis in micro-econometrics.

Occasionally, real, randomized experiments do occur in economics. One of the most famous examples is the negative income tax experiment, performed in New Jersey and a few other locations in the USA (a discusssion of this and other experiments is given by Ferber and Hirsch, 1982). There is also a growing literature on economic experiments in game theory and individual decision making, but those experiments, which are frequently very revealing, rarely have an elaborate econometric follow up (see Kagel and Roth, 1995). There seems to be an empirical law, stating that the more advanced the econometrics used, the less informative the empirical result.

Even where experiments are feasible, experimental design may be an intricate issue. One of the problems of such experiments is that the initiation of an experiment may influence the outcomes (this is just one example where the observer may influence the subject). An alternative to experimental design is the use of 'indirect randomization' by means of so-called natural experiments. An example is the Vietnam draft lottery which was based on birthdays of young male adults. The lottery can be used as an indirect randomization device to study the effects of veteran status on lifetime earnings (Angrist, 1990). The great advantage of 'natural experiments' is that problems of selection bias are strongly reduced. The drawback is that natural experiments are rare.

Hence, in practice econometricians mostly have to do without experiments. From the early days on, they tried to provide stand-ins for real experimentation. First, it was argued that economic models themselves could substitute as alternatives to experiments. However, this was not intended so much as to justify probabilistic inference, but as, primarily, to level economics with the 'experimental' sciences. Then Haavelmo introduced his version of experimental design in order to make economic theories (statistically) operational. Again, this could not provide the justification for using the probabilistic framework. The founders of econometrics tried to adapt the sampling approach to a non-experimental small sample domain. They tried to justify this with *a priori* and analytical arguments. However, the ultimate argument for a 'probability approach in econometrics' consists of a mixture of metaphors, metaphysics and a pinch of bluff.

This failure has two sides. The first is that it is not necessarily harmful. One can fruitfully slash a hammer without understanding Newtonian mechanics. Similarly, one can do useful data analysis and empirical inference with statistical techniques which are not perfectly understood and which may even be formally inappropriate. There is however another side. The approach which Haavelmo developed and which the Cowles Commission further refined took econometrics away from the pragmatic approach which characterizes British empiricism. Analytical fundamentalism led to an emphasis on solving relatively minor issues (such as simultaneous equations bias) at the cost of neglecting the empirically serious questions related to inference in a non-experimental setting (see also the complaint of Modigliani, presented in chapter 7, section 2, below). The false probabilistic justification did not help econometrics, when in the early 1970s applied econometrics became less and less respected. The following chapter deals with the Cowles tradition and some recent alternative econometric methodologies that were inspired by the growing discomfort with Haavelmo's (1944) 'probability approach in econometrics'.

Notes

1. Note, however, that Bayesian parametric inference also involves sampling assumptions which have to be justified.
2. The prewar history of econometrics is well described in Morgan (1990). She does not elaborate on the various interpretations of probability theory that underlie the work of these authors and their successors. In this sense, the following discussion complements her work. Another useful reference is Hendry and Morgan (1995), which contains a number of 'classic' papers in early econometrics.
3. Pearson reported no effects of parents' use of alcohol on their children's health. Keynes objected and argued that Pearson's results were nearly worthless: Pearson's samples were not representative for the whole population and the homogeneity of the population was questioned. Skidelski (1983, pp. 223–6) suggests that Keynes' criticism of Pearson resulted as much from his Victorian disagreement with Pearson's conclusions, as from legitimate statistical criticism.
4. Of course, least squares is just one tool that can be applied in problems of estimation. Other criteria (e.g. the minimum absolute deviation estimator), and other methods, such as the method of moments (due to Karl Pearson) and the likelihood approach (Fisher's maximum likelihood, Jeffreys' Bayesianism) are used as well.
5. Granger and Newbold (1974) diverge from this terminology, by using the phrase spurious regression where Yule used the word nonsense.
6. An alternative to the justification of sampling assumptions in parametric econometrics can be found in distribution-free methods, non-parametric econometrics and method of moments estimation.
7. Reference is made to the fifth edition of Yule's *Introduction to the Theory of Statistics*. Note that independence follows from von Mises' randomness condition, a point not considered by Mills.
8. Note that, unlike Crum, Patton and Tebbutt, Mills (1938, pp. 460–4) gives a fairly correct interpretation of Fisher's sample design.
9. Tintner (1940, p. 25) justifies inference from the sample to the '(hypothetically infinite) population' by means of the law of large numbers. He does not discuss when a sample of economic data may be considered 'large'.
10. It should be noted however that the theory of experimental design was first presented in a reference book format by Fisher ([1935] 1966) – five years after the appearance of the first edition of Ezekiel's book.
11. Ezekiel (1941, p. 350) points out that the theory of simple sampling 'assumes that each observation in a sample is selected purely at random from all the terms in the original universe. It also assumes that successive samples are selected in such a way that values found in one sample have no relation or

connection with the values found in the next sample.' To this sentence, a footnote is attached which refers to the English 1939 edition of von Mises ([1928] 1981).

12. Tinbergen (1936), which was written in Dutch, uses the English noun 'sample' instead of the familiar Dutch 'steekproef'. Tinbergen (1939a, p. 28) mentions the notions of 'sample' and the 'universe' from which it is drawn, but without further discussion.
13. See the quotation of Fisher (1956, pp. 99–100) in chapter 3, section 4.3: 'To regard the test as one of a series is artificial; . . . what is really irrelevant, [is] the frequency of events in an endless series of repeated trials which will never take place'.
14. Haavelmo may have been inspired by a similar change of wording, made by Neyman and Pearson. In 1928, they write '[w]e may term Hypothesis A the hypothesis that the population from which the sample Σ has been randomly drawn is that specified, namely Π' (Neyman and Pearson, 1967, p. 1). It evolves in 1931 to '[t]he set Ω of admissible populations π_t, from which the Σ_t may have been drawn' (p. 118). In 1933, the sample point E and sample space W are added to their terminology.
15. See also *ibid.*, p. 6 and pp. 13–14 where a design of experiments is related to the 'crucial experiment' in physics.

7 Econometric modelling

> It takes a lot of self-discipline not to exaggerate the probabilities you would have attached to hypotheses before they were suggested to you.
>
> L. J. Savage (1962)

1 Introduction

The frequency interpretation of probability presumes that the statistical model is given by underlying theory, or by experimental design (the 'logic of the laboratory'). The task of the statistician is to estimate the parameters and to test hypotheses.

Econometrics is different. The underlying economic theories to justify a model are not well established. Hence, part of the inference concerns the theory and specification of the model. The usually unique, non-experimental data (sample) are used for specification purposes, the specification thus derived is used for the purpose of inference. There is no experimental design to justify distributional assumptions: *if* they are investigated, the same data are used for this purpose. The pioneer of econometrics, Tinbergen (1939b, p. 6), had already recognized that it is not possible to work with a sequence of two separate analytical stages, 'first, an analysis of theories, and secondly, a statistical testing of those theories'. Modelling turned out to become an Achilles' heel for econometrics.

Moreover, one task of inference is theoretical progress, hence, the search for new specifications. The probability approach, advocated by Haavelmo and his colleagues at the Cowles Commission, is not suitable for inference on rival theories and specifications in a situation of unique, non-experimental data. Section 2 deals with the Cowles methodology, which long dominated 'textbook econometrics' but has became increasingly less respected.

Specification uncertainty has provoked at least four alternative approaches to econometric modelling which will be discussed in this chapter. They are:

- 'profligate' modelling based on Vector Autoregressions, which rejects 'incredible' over-identifying restrictions (Sims; discussed in section 3)

- reductionism, emphasizing diagnostic testing (Hendry; section 4)
- sensitivity analysis which quantifies the effect of specification searches on parameters of interest (Friedman, Leamer; section 5)
- calibration, or structural econometrics without probability (Kydland, Prescott; section 6).

The order of the methodologies is by the increasing emphasis given by each to economic theory. Note that the methodologies presented here do not exhaust all methods of inference in economics. A serious contender is experimental economics, which is becoming increasingly popular. It can be regarded as a complement to econometrics, where the Fisher methodology is, in principle, applicable (some merits of experimental economics are discussed in chapter 9). The merits of maximum likelihood versus method of moments and non- (or semi-)parametric methods are not discussed here: I regard them as rival techniques, not methodologies. Co-integration, which is a very important technique in modern econometrics, again is not considered as another methodological response to specification uncertainty (unlike Darnell and Evans, 1990).

2 The legacy of Koopmans and Haavelmo to econometrics

2.1 An experimental method?

Haavelmo and the other contributors to what now is known as the 'Cowles approach to econometrics' deserve praise for their efforts to provide methodological foundations for econometrics. Haavelmo views science in the following way (1944, p. 12): in the beginning, there is 'Man's craving for "explanations" of "curious happenings"'. There is not yet systematic observation, the explanations proposed are often of a metaphysical type. The next stage is the '"cold-blooded" empiricists', who collect and classify data. The empiricists recognize an order or system in the behaviour of the phenomena. A system of relationships is formulated to mimic the data. This model becomes more and more complex, and needs more and more exceptions or special cases (perhaps Haavelmo has the Ptolemaic system in mind), until the time comes to start reasoning (the Copernican revolution?). This '*a priori reasoning*' (a somewhat odd notion, in light of Haavelmo's starting point of collecting data) leads to 'very general – and often very simple – principles and relationships, from which *whole classes* of apparently very different things may be deduced'. This results in hypothetical constructions and deductions, which are very fruitful if, first, there are in fact 'laws of Nature', and, secondly, if the method employed is efficient (it goes without saying that the next step of Haavelmo is to recommend the probabilistic approach to inference). Finally, Haavelmo (p. 14) summarizes his view by quoting Bertrand Russell favourably: 'The actual procedure of science consists of an alternation of observation, hypothesis, experiment and theory.'

This view of the development of science may be a caricature, but not an uncommon one, certainly not around 1944. It suggests that science develops from an initial ultra-Baconian phase to the hypothetico-deductive method that became popular among the logical positivists in the 1940s (see e.g. Caldwell, 1982, pp. 23–7). The so-called correspondence rules that relate theoretical entities to observational entities in logical positivism can be found in Haavelmo's emphasis on experimental design. Despite the similarity between this view of Haavelmo on scientific development, and the view of contemporaneous philosophers, it is unlikely that Haavelmo really tried to apply the logical positivist doctrine to econometrics. In most cases, his ultimate argument is the fruitfulness of his approach, not its philosophical coherency.

Haavelmo regards econometrics as a means of inference for establishing phenomenological laws in economics. Without referring to Keynes, he (1944, p. 12) rejects the assertion that intrinsic instability of economic phenomena invalidates such an approach: 'a phrase such as "In economic life there are no constant laws", is not only too pessimistic, it also seems meaningless. At any rate, it cannot be tested' (Haavelmo, 1944, p. 12). Economists should try to follow natural scientists, who have been successful in choosing 'fruitful ways of looking upon physical reality', in order to find the elements of invariance in economic life (p. 13; see also pp. 15–16). Haavelmo (pp. 8–9) considers a model as

> an *a priori hypothesis* about real phenomena, stating that every system of values that we might observe of the 'true' variables will be one that belongs to the set of value-systems that is admissible within the model... Hypotheses in the above sense are thus the joint implications – and the only testable implications, as far as observations are concerned – of a theory *and* a design of experiments. It is then natural to adopt the convention that a theory is called true or false according as the hypotheses implied are true or false, when tested against the data chosen as the 'true' variables. Then we may speak, interchangeably, about testing hypotheses or testing theories.

This suggests a one to one relation between hypotheses or models and theories. However, economics models are not instances of economic theories (Klant, 1990). Haavelmo's argument is, therefore, invalid.

Haavelmo rejects a Baconian view of scientific inference: it is not permissible to enlarge the 'model space' (the set of *a priori* admissible hypotheses) after looking at the data. What is permissible, is to start with a relatively large model space and select the hypothesis that has the best fit (1944, p. 83). But, if we return to Keynes' complaints, what if the *a priori* model space is extremely large (because there is no way to constrain the independent variation in a social science, i.e. too many causes are at work and moreover, these causes are unstable)? Haavelmo has no clear answer to this question. He holds that 'a vast amount of experience' suggests that there is much invariance (autonomy) around, but does not substantiate this claim. Neither does he resort to Fisher's solution to this problem, randomization.

Haavelmo is more 'liberal' than Koopmans, whose insistence on the primacy of economic theory has become famous (see in particular his debate with Vining). Koopmans (1937, p. 58) had already argued that the significance of statistical results depends entirely on the validity of the underlying economic analysis, and asks the statistician

> for a test affirming the validity of the economist's supposed complete set of determining variables on purely statistical evidence [which] would mean, indeed, [turning] the matter upside down.

The statistician may be able, in some cases, to inform the economist that he is wrong, if the residuals of an equation do not have the desired property. Haavelmo adds to this stricture that the statistician may formulate a theory 'by looking at the data' – the subject of the next sections of this chapter.

What can be said about the effectiveness of Haavelmo's programme for econometrics? Some argue that it revolutionized econometrics. Morgan (1990, p. 258) claims:

> By laying out a framework in which decisions could be made about which theories are supported by data and which are not, Haavelmo provided an adequate experimental method for economics.

The word 'experimental' has been abused by Haavelmo and used in an equally misleading manner by Morgan. The revolution, moreover, may be regarded as a mixed blessing: the probability approach to econometrics has led to a one-sided and often unhelpful emphasis on a limited set of dogmas. Modigliani recalls (in Gans and Shepherd, 1994, p. 168) how one of his most influential papers was rejected by Haavelmo, then editor of *Econometrica*. This was the paper which introduced the Duesenberry–Modigliani consumption function. Haavelmo's argument was that the important issue at the time was not inventing new economic hypotheses, but improving estimation methods for dealing with simultaneity. Those were needed to sophisticate the structural econometrics programme which took a Walrasian set of simultaneous equilibrium equations as a given starting point for the econometric exercise of estimating the values of the 'deep parameters' involved.

2.2 Structure and reduced form

The most important feature of the Cowles programme in econometrics is the just mentioned 'structural' approach. This takes the (romantic) approach that, in economics, everything depends on everything. Economic relations are part of a Walrasian general equilibrium system. Estimating single equations may lead to wrong inferences if this is neglected.

All introductory econometric textbooks spend a good deal of space on this problem. An elementary example is consumer behaviour, where consumer demand c is a function of a price p and income y. If supply also depends on those variables, then the result is a system of equations. In general terms, such a system can be represented by endogenous variables $\mathbf{y} = y_1, y_2, \ldots, y_m$, exogenous variables $\mathbf{x} = x_1, x_2, \ldots x_n$, and disturbances u. The system can be written as:

$$\begin{aligned} \gamma_{11}y_{1t} + \gamma_{21}y_{2t} + \ldots + \gamma_{m1}y_{mt} + \beta_{11}x_{1t} + \ldots + \beta_{n1}x_{nt} &= u_{1t} \\ \gamma_{12}y_{1t} + \gamma_{22}y_{2t} + \ldots + \gamma_{m2}y_{mt} + \beta_{12}x_{1t} + \ldots + \beta_{n2}x_{nt} &= u_{2t} \\ \gamma_{1m}y_{1t} + \gamma_{2m}y_{2t} + \ldots + \gamma_{mm}y_{mt} + \beta_{1m}x_{1t} + \ldots + \beta_{nm}x_{nt} &= u_{mt} \end{aligned} \tag{1}$$

or, in matrix notation,

$$\mathbf{y}_t'\Gamma + \mathbf{x}_t'\mathrm{B} = \mathbf{u}_t'. \tag{2}$$

In this case of a system of equations, a regression of the variables in the first equation of the system does not yield meaningful information on the parameters of this equation. One might solve the system of equations for $\mathbf{y}_t$ (given that Γ is invertible):

$$\mathbf{y}_t' = -\mathbf{x}_t'\mathrm{B}\Gamma^{-1} + \mathbf{u}_t'\Gamma^{-1}, \tag{3}$$

in which case regression of $\mathbf{y}_t$ on $\mathbf{x}_t$ does yield information on the reduced form regression coefficients BC^{-1}, but not on the separate parameters of interest in the structural equations.

For such information, one needs to satisfy the identification conditions. The most well-known type of condition is the (necessary, but not sufficient) order condition, which says that the number of excluded variables from equation j must be at least as large as the number of endogenous variables in that equation.[1] The choice of the exclusion restrictions may be inspired by economic theory. The Cowles methodology is based on the valid provision of such restrictions.

Equation (3) is useful to obtain the reduced form parameters $\Pi = B\Gamma^{-1}$. In very specific circumstances (exact identification of an equation), the reduced form ordinary least squares regression can be used for consistent estimation of the structural parameters (this is known as indirect least squares). In general, however, this is not possible and more advanced techniques are needed. The favoured method of the Cowles econometricians is full information maximum likelihood. Because of its computational complexity (and also sensitivity to specification error), other methods have been proposed for estimation equations with endogenous regressors. Two stage least squares (due to Theil) or the method of instrumental variables is the most popular of those methods.

2.3 Theory and practice

The degree to which the ideas of the Cowles Commission really have been practised in applied econometrics is limited: in theory, it was widely endorsed, in practice, econometricians tried to be pragmatic. There are a number of reasons for this deviation. First, the Neyman–Pearson methodology does not apply without repeated sampling. Second, the Cowles Commission emphasized errors in equations (simultaneous equations bias) at the cost of errors in variables (ironically the main topic of Koopmans, 1937). In empirical investigations, simultaneous equations bias appears to be less important than other sources of bias, moreover, methods to deal with simultaneity (in particular full information maximum likelihood) are either less robust than single equation methods, and/or difficult to compute (see Fox, 1989, pp. 60–1; Gilbert, 1991b).

Third, the view that the theory (or 'model space') should be specified beforehand is not very fruitful. It is hardly avoidable that the distinction between model construction and model evaluation is violated in empirical econometrics (Leamer, 1978; Heckman, 1992; Eichenbaum, 1995). It is also not desirable. I concur with Eichenbaum (p. 1619), who argues that the programme has proven 'irrelevant' for inductive aims in economics. If the set of admissible theories has to be given first (by economic theory) before the econometrician can start his work of estimation and testing, then data can do little more than fill in the blank (i.e. parameter values). In practice, there is *and should be* a strong interference between theory and empirical models, as this is the prime source of scientific progress.

Only recently, the Cowles Commission sermon seems to have gained practical relevance, due to the new-classical search for 'deep parameters', continued in real business cycle analysis (see the discussion on calibration, section 6). This approach may be accused of overshooting in its faith in economic theory, which leads to a 'scientific illusion' (Summers, 1991). The criticism of Summers is of interest, not because it is philosophically very sound (it isn't), but because it shows some weaknesses of formal econometrics in the hypothetical-deductive tradition: in particular, the lack of empirical success and practical utility.[2] In a nutshell, Summers (following Leamer) argues for more sensitivity analysis, exploratory data analysis and other ways of 'data-mining' that are rejected by the Cowles econometricians.

The work of the early econometricians relies on a blend of the views of von Mises, Fisher and Neyman–Pearson. Von Mises' fundamental justification of the frequency theory was endorsed without taking the imperative of large samples seriously. Fisher, on the other hand, did not hesitate to use statistics for small sample analysis. This helped Fisher's popularity among econometricians, although Fisher had little interest in econometrics. His theory of experimental design, which he considered crucial, has been ignored in econometrics until very recently. Similarly, the Neyman–Pearson method of quality control in repeated samples is hard to reconcile with econometric practice. If there ever was a 'probability revolution' in econometrics (as Morgan, 1990, holds), it was based on an arguably deliberate misconception of prevailing probability theory. It yielded Haavelmo a Nobel memorial prize for economics in 1989.

3 Let the data speak

3.1 Incredible restrictions

Liu (1960) formulated a devastating critique of the structural econometrics programme of the Cowles Commission. Consider the following investment equation to illustrate the point (p. 858, n. 8), where y is investment, x_1 profits, x_2 capital stock, x_3 liquid assets, and x_4 the interest rate:

$$y = a_0 + a_1x_1 + a_2x_2 + a_3x_3 + a_4x_4 + u. \tag{4}$$

Now suppose that the estimates for a_2 and a_4 yield 'wrong' signs, after which x_4 is dropped. The new regression equation is:

$$y = a_0' + a_1'x_1 + a_2'x_2 + a_3'x_3 + u'. \tag{5}$$

However, unknown to the investigator, the 'true' investment equation is a more complex one, for example:

$$y = b_0 + b_1x_1 + b_2x_2 + b_3x_3 + b_4x_4 + b_5x_5 + u^*. \tag{6}$$

Textbook econometrics teaches that the expected values of the vectors of parameters $\mathbf{a}$ and $\mathbf{a}'$ are functions of the 'true' parameters vector, $\mathbf{b}$. Those functions depend on the size of the parameters and the correlation between the omitted and included regressors. The simplification from (4) to (5) may be exactly the wrong strategy, even if the deleted variable is insignificant in (4). Liu concludes that over-identifying restrictions in structural econometrics are very likely to be artifacts of a specification search and not justifiable on theoretical grounds. The investment equation is not identified, despite the apparent fulfilment of the identifying conditions. Liu's advice is to concentrate on reduced form equations, particularly if prediction is the purpose of modelling.[3] His scepticism on parametric identification remained largely ignored, until Sims (1980) revived it forcefully.

Sims title, 'Macroeconomics and reality', mimics the title of chapter 5 of Robbins' ([1932] 1935), 'Economic fluctuations and reality'. Both authors are sceptical of prevailing econometric studies of macro-economic fluctuations. Whereas Robbins rejects econometrics altogether, Sims rejects econometric practice and suggests an alternative econometric methodology.

Sims (1980, p. 1) observes that large-scale statistical macro-economic models 'ought to be the arena within which macroeconomic theories confront reality and thereby each other'. But macro-econometric models lost much support since the 1970s. The reason, Sims argues, is the same one as Liu had already indicated: the models are mis-specified and suffer from incredible identifying assumptions. As an alternative, he proposes to 'let the data speak for themselves' by reducing the burden of maintained hypotheses imposed in traditional (structural, Cowles-type) econometric modelling. This is the most purely inductivist approach taken in econometrics, a Baconian extension of the Box–Jenkins time series methodology (with its roots in earlier work of Herman Wold). The difference from Liu and Box–Jenkins methods is that Sims proposes to respect the interdependent nature of economic data by means of a system of equations.

This is done by introducing an unrestricted Vector Autoregression (VAR), in which a vector of variables $\mathbf{y}$ is regressed on its past realizations:

$$A(L)\mathbf{y}_t = \mathbf{u}_t, \tag{7}$$

where L denotes the lag operator. This is also known as an unrestricted reduced form of a system of equations. Parametric assumptions are needed to make (7) operational. Sims (1980) specifies it as a linear model. This assumption is not innocuous and has been rightly criticized by Leamer (1985). Non-linearities may be important if policy ventures outside the historical bandwidth (the kinds of 'historical events' that Friedman uses for non-statistical identification of causes of business cycle phenomena). A VAR approach may easily underestimate the importance of such historical events (but the same criticism applies to the mainstream approach criticized by Sims as well as to Leamer's own empirical work).

The linear version of the VAR for the n-dimensional vector $\mathbf{y}$ can be formulated as

$$\mathbf{y}_t = \sum_{i=1}^{L} A_i \mathbf{y}_{t-i} + \mathbf{u}_t, \tag{8}$$

$$\text{with } \mathcal{E}(\mathbf{u}_t \mathbf{u}'_{t-i}) = \Sigma \text{ if } i = 0$$
$$= 0 \text{ otherwise.}$$

For simplicity, the variance–covariance matrix is assumed to be constant over time. In theory, the lag order L may be infinite. If the vector stochastic process $\mathbf{y}_t$ is stationary, it can be approximated by a finite order VAR. If $\mathbf{u}_t$ follows a white-noise process, the VAR can be estimated consistently and efficiently by OLS regression (Judge *et al.*, [1980] 1985, p. 680).

3.2 Profligate modelling

The VAR approach is radically different from the plea of Hendry (discussed below), to 'test, test and test'. VAR econometricians are rarely interested in significance tests, apart from testing for the appropriate lag length (by means of a likelihood ratio test or an information criterion) and 'Granger causality' tests. Lag length indeed is an important issue. The VAR contains n^2L free parameters to be estimated. A modest VAR for five variables with five lags yields 125 parameters. Estimating unrestricted VARs easily results in over-fitting. This is at odds with the theory of simplicity presented in chapter 5. Instead of parsimonious modelling

Sims (1980, p. 15) favours a strategy for estimating 'profligately'. This makes predictions less reliable, as the model does not separate 'signal' from 'noise'. Hence, unrestricted VARs (such as in Sims, 1980) tend to be sample-dependent (idiosyncratic) and are, therefore, of limited interest. Zellner rightfully calls such models 'Very Awful Regressions'. Further reduction or adding prior information is desirable for obtaining meaningful results.

The latter option is discussed in Doan, Litterman and Sims (1984), who advocate a Bayesian view on forecasting. These authors argue that exclusion restrictions (setting particular parameters *a priori* equal to zero) reflect an unjustified degree of absolute certainty, while prior information on behalf of the retained variables is not incorporated. Hence, the traditional approach to structural econometrics is a mixture of imposing too much and too little prior information. Bayesian VARs may provide a useful compromise. However, specifying an appropriate prior distribution for a complicated simultaneous system is hard, as, for example, is revealed in Doan, Litterman and Sims (1984). This has hampered the growth of the VAR research programme. A second problem that impeded the programme was the return of incredible restrictions via the back door. I will turn to this issue now.

3.3 Innovation accounting

In order to explain macro-economic relationships, proponents of the VAR methodology decompose variances of the system to analyse the effect of innovations to the system. If there is a surprise change in one variable, this will have an effect on the realizations of its own future as well as on those of other variables both in the present and the future. This can be simulated by means of impulse response functions. The moving average representation of the VAR is:

$$y_t = \sum_{i=1}^{L} B_i \mathbf{u}_{t-i}. \tag{9}$$

For 'innovation accounting', as it is called, one needs to know how an impulse via $\mathbf{u}_t$ affects the future of the variables in $\mathbf{y}$. A problem is that there are many equivalent representations of (9) (as one may replace B_i by B_iG and $\mathbf{u}_{t-i}$ by $G^{-1}\mathbf{u}_{t-i}$), hence many different responses of the system to an impulse are conceivable. One may choose G such that B_0G is the identity matrix $\mathbf{I}$. Then, the innovations are non-orthogonal if the covariance matrix Σ is not diagonal (see Doan, 1988, p. 8.8, for a discussion of orthogonalization; the following is based on this source). In this non-diagonal case the innovations are correlated. Hence, simulating the responses to a single impulse in one variable will result in an interpretation problem as, historically, such separate impulses do not occur. For this purpose, VAR modellers orthogonalize Σ such that $G^{-1}\Sigma G'^{-1} = \mathbf{I}$. As a result, the orthogonalized innovations $\mathbf{v}_t = \mathbf{u}_t G^{-1}$ (with $\mathcal{E}(\mathbf{v}_t \mathbf{v}_t') = \mathbf{I}$) are uncorrelated across time and across equations. Impulse accounting can begin without this interpretation problem.

But a subsequent problem arises, which is the major weakness of the innovation accounting approach: how to choose G. The Choleski factorization (G lower triangular with positive elements on the diagonal) is one alternative. It has the attractive property that the transformed innovations can be interpreted as impulses of one standard error in an ordinary least squares regression of variable j in the system on variables 1 to $j-1$. However, there are as many different Choleski factorizations as orderings of variables (120 in a system of five variables). The effect of impulses is crucially dependent on the choice of ordering. This choice involves the same incredible assumptions as those that were criticized to start with. The ordering should be justified, and the sensitivity of the results to changes in the ordering must be examined. Interpreting the results may become a hard exercise.

3.4 VAR and inference

The VAR methodology intends to reduce the scope for data-mining by its choice to include all variables available. But even the VAR allows options for data-mining:

- the choice of variables to begin with
- the choice of lag length
- the choice of the error decomposition to use.

For these reasons, the VAR does not seem the right response to specification uncertainty and data-mining. Economic theory remains an indispensable tool for deriving useful empirical models. This does not mean that the VAR is without merits. Two branches of empirical research should be mentioned. The first is due to Sims' own work, the second is related to the analysis of consumption and the permanent income theory. A third line of research which owes much to the VAR methodology is co-integration which will not be discussed here.

Sims (1996, p. 117) provides a brief summary of useful findings about economic regularities related to the role of money and interest in business cycles. This research line is presented as an extension of the monetary analysis of Friedman and Schwartz. The initial finding that 'money

(Granger) causes income' (presented by Friedman and Schwartz and confirmed by Sims, 1972) was re-adjusted when a VAR exercise showed that short-term interest innovations accounted for most of the variation in output and 'absorbed' the direct effect of money (Sims, 1980). An extensive research line on the interpretation of interest innovations as policy variables resulted, including institutional studies. Sims believes that this has resulted in a number of well-established 'stylized facts' which turned out to be stable across a range of economies. If this is correct, then the VAR-methodology can claim an important accomplishment. However, Sims had already complained that other researchers (in particular real business cycle analysts) do not accept these stylized facts in their own research. Might the explanation be that VARs are still considered as too idiosyncratic, sample-dependent and difficult to interpret from an economic theorist's perspective? An estimated elasticity yields clear information, whereas an impulse response picture is the miraculous outcome of a mysterious model.

The second line of research in which VARs have indeed been enormously helpful is the analysis of consumption, in particular the permanent income consumption function (see e.g. Campbell, 1987). Ironically, this line of research has a strong theoretical motivation, at odds with the much more inductive intentions originally motivating the VAR methodology. It has a highly simplified structure (e.g. a two-variable VAR). The results are directly interpretable in terms of economic theory. It has yielded a number of new theoretical research lines, like those on liquidity constraints and excess sensitivity of consumption to income. The results have reached the textbook level of macro-economics and will have a lasting impact on macro-economic research.

In sum, time-series methods are most useful when economic theory can be related to the statistical model. This is the case in the examples mentioned above. It also is a feature of some alternative approaches that deserve to be mentioned, such as co-integration methods (which grew out of the work of Granger) and the 'structural econometric modelling time series analysis' of Palm and Zellner (1974). But it is an illusion that econometric inference can be successful by letting only data speak.

4 The statistical theory of reduction

4.1 Reduction and experimental design

For Fisher, the 'reduction of data' is the major task of statistics. The tool to achieve this goal is maximum likelihood: 'among Consistent Estimates, when properly defined, that which conserves the greatest amount of information is the estimate of Maximum Likelihood' (quoted on p. 44 above). One of the most important of Fisher's criteria is sufficiency, 'a statistic of this class alone includes the whole of the relevant information which the observations contain' (Fisher, [1925] 1973, p. 15).

Fisher's statistical theory of reduction is not only based on his method of maximum likelihood, but also on his theory of experimental design. This is a necessary requirement in order to asses 'the validity of the estimates of error used in tests of significance', something that 'was for long ignored, and is still often overlooked in practice' (Fisher, [1935] 1966, p. 42). The crucial element in his theory of experimental design is randomization (see chapters 3, section 3.3 and 6, section 3.4 above).

Randomization is Fisher's physical means for ensuring normality or other 'maintained hypotheses' of a statistical model. Student's *t*-test for testing the null hypothesis that two samples are drawn from the same normally distributed population is of particular interest for Fisher (see also 'Gosset' in Personalia). He regards it as one of the most useful tests available to research workers, 'because the unique properties of the normal distribution make it alone suitable for general application.' (Fisher, [1935] 1966, p. 45). Fisher's disagreement with statisticians of the 'mathematical school' (in particular Neyman) results largely from different opinions about the effect of randomization on normality. According to Fisher, proper randomization justifies the assumption of normality and the use of parametric tests. Assuming normality is not a drawback of tests of significance. Research workers can devise their experiments in such a way as to validate the statistical methods based on the Gaussian theory of errors. The experimenters 'should remember that they and their colleagues usually know more about the kind of material they are dealing with than do the authors of textbooks written without such personal knowledge' (p. 49).

4.2 The Data Generation Process[4]

Fisher's theory of experimental design is unfit for analysis of non-experimental data. The metaphor 'experiment' has remained popular in econometrics (see Haavelmo, 1944; Florens, Mouchart and Rolin, 1990[5]) but is rarely related to the kind of experiment that Fisher had in mind. Hence, Fisher's methods of reduction usually do not apply to econometrics.

An alternative, that still owes much to (but rarely cites) Fisher is the theory of reduction and model design. This is sometimes called 'LSE-econometrics' but statistical reductionism, or simply reductionism, seems more appropriate. David Hendry is the most active econometrician in this school of thought.

Reductionism starts from the notion of a ‘Data Generation Process’ (DGP). It is a highly dimensional probability distribution for a huge vector of variables, $\mathbf{w}$,

$$DGP \equiv f(\mathbf{W}_T^1 | \mathbf{W}_0, \omega_T^1) = \prod_{t=1}^{T} f(\mathbf{w}_t | \mathbf{W}_{t-1}^1, \omega_t), \quad (10)$$

conditional on initial conditions ($\mathbf{W}_0$), parameters $\omega_t \in \Omega_t$, continuity (needed for a density representation) and time homogeneity ($f_t(\cdot) = f(\cdot)$). A DGP is a well-defined notion in Monte Carlo studies, where the investigator is able to generate the data and knows exactly the characteristics of the generation process. However, reductionists transfer it to applied econometrics, which is supposed to deal with the study of the properties of the DGP. The intention of theory of reduction is to bring this incomprehensive distribution down to a parsimonious model, without loss of relevant information. Marginalizing and conditioning are two key notions of this theory of reduction.

In applied econometrics, Hendry (1993, p. 73) argues, ‘the *data* are given, and so the distributions of the dependent variables are already fixed by what the data generating process created them to be – I knew that from Monte Carlo’. Apart from the data, there are models (sets of hypotheses). Given the data, the goal is to make inferences concerning the hypotheses. In the theory of reduction, the idea is that the DGP (or population) can be used as a starting point as well as a goal of inference. Econometric models are regarded as reductions of the DGP (p. 247).

Although the DGP is sometimes presented as ‘hypothetical’ (e.g. p. 364), there is a tendency in Hendry’s writings to view the DGP not as an hypothesis (one of many possible hypotheses), but as fact or reality. Sometimes the DGP is presented as ‘the relevant data generation process’ (p. 74). In other instances, the DGP becomes the ‘actual mechanism that generates the data’ (Hendry and Ericsson, 1991, p. 18), or simply ‘the actual DGP’ (Spanos, 1986). The DGP is reality and a model of reality at the same time.[6] Philosophers call this ‘reification’. Once this position is taken, weird consequences follow. Consider the ‘information taxonomy’ that ‘follows directly from the theory of reduction given the sequence of steps needed to derive any model from the data generation process’ (Hendry, 1993, pp. 271, 358). Or, econometric models which are ‘derived and *derivable* from the DGP’ (Hendry and Ericsson, 1991, p. 20; emphasis added). Consider another example of reification: the proposition that ‘it is a common empirical finding that DGPs are subject to interventions affecting some of their parameters’ (p. 373). Rather, those empirical findings relate to statistical models of the data, not to the DGP or its parameters. Paraphrasing de Finetti, THE DGP DOES NOT EXIST. It is an invention of our mind.

Models are useful approximations to reality. Different purposes require different models (and often different data, concerning levels of aggregation, numbers of observations). The idea that there is one DGP waiting to be discovered is a form of 'scientific realism' (see chapter 9, section 2 for further discussion of realism). This view entails that the task of the econometrician is 'to model the main features of *the* data generation process' (Hendry, 1993, p. 445; emphasis added). Instrumentalists do not have a similar confidence in their relationship to the truth. They regard models as useful tools, sometimes even intentionally designed to *change* the economic process. This may involve self-fulfilling (or self-denying) prophecies, which are hard (if not impossible) to reconcile with the notion of a DGP.

This is not to deny that there may be many valid statistical models even in absence of 'the' DGP. Statistical theory does not depend on such a hypothesis. A time series statistician may formulate a valid univariate model (see Chung, 1974, for necessary conditions for the existence of a valid characterization), while a Cowles econometrician will usually design a very different model of the same endogenous variable. Although both may start with the notion of a DGP, it is unlikely that these hypothesized DGPs coincide. Their DGPs depend on the purpose of modelling, as do their reductions.

4.3 General to specific

An important characteristic of the reductionist methodology is the general-to-specific approach to modelling. It was initially inspired by Denis Sargan's Common Factor test, where a general model is needed as a starting point for testing a lagged set of variables corresponding to an autoregressive error process. Another source of inspiration was the reductionist approach developed at CORE (in particular, by Jean-Pierre Florens, Michel Mouchart and Jean-François Richard). The theory of reduction deals with the conditions for valid marginalizing and conditioning (relying on sufficient statistics). However, note that Fisher does not claim to reduce the DGP, but rather to reduce a large set of data without losing relevant information.

The basic idea of general-to-specific modelling is that there is a starting point for all inference, the data generation process. All models are reductions of the DGP, but not all of them are valid reductions. In the following, a brief outline of reduction is sketched. Subsequently, a methodological critique is given. Reconsider the DGP, (10). The variable vector **w** contains only few variables that are likely to be relevant. These are labelled $\mathbf{y}^*$, the remainder is $\mathbf{w}^*$ (nuisance variables). Using the definition of conditional probability,[7] and assuming parameter constancy for simplicity, one can partition the DGP:

$$f(\mathbf{w}_t^*, \mathbf{y}_t^* | \mathbf{W}_{t-1}^*, \mathbf{Y}_{t-1}^*, \omega_1) = f_1(\mathbf{y}_t^* | \mathbf{W}_{t-1}^*, \mathbf{Y}_{t-1}^*, \omega_1) \\ f_2(\mathbf{w}_t^* | \mathbf{W}_{t-1}^*, \mathbf{y}_t^*, \mathbf{Y}_{t-1}^*, \omega_z) \tag{11}$$

where $\mathbf{W}_{t-1}^* = (\mathbf{W}_{t-1}^1)^*$. The interest is on the marginal density, f_1. If ω_1 and ω_2 are variation free ($(\omega_1, \omega_2) \in \Omega_1 \times \Omega_2$; i.e. f_2 does not provide information about Ω_1) then, a 'sequential cut' may be operated: consider only f_1. Assuming furthermore that $\mathbf{y}_t^*$ and $\mathbf{W}_{t-1}^*$ are independent conditionally on $\mathbf{Y}_{t-1}^*$ and ω_1, the nuisance variables can be dropped from f_1, which yields:

$$f(\mathbf{y}_t^* | \mathbf{Y}_{t-1}^*, \omega_1). \tag{12}$$

This can be represented using a specific probabilistic and functional form (e.g. a VAR). Note that this marginalizing stage of the reduction process proceeds entirely implicitly. It is impossible to consider empirically all conceivable variables of interest to start with. Note also that many hidden assumptions are made (e.g. going from distributions to densities, assuming linearity, parametric distributions, constant parameters). Some of them may be testable but most are taken for granted.[8]

The second stage, conditioning, aims at a further reduction by introducing exogeneity into the model. For this purpose, decompose $\mathbf{y}^*$ into **y** (endogenous) and **x** (exogenous) variables. Rewrite (12) as:

$$f(y_t^* | \mathbf{Y}_{t-1}^*, \omega_1) = f(y_t | \mathbf{X}_{t-1}, x_t, \mathbf{Y}_{t-1}, \theta_1) f(x_t | \mathbf{X}_{t-1}, \mathbf{Y}_{t-1}, \theta_2) \tag{13}$$

where θ_1 are the parameters of interest. If x_t is 'weakly exogenous' (θ_1 and θ_2 are variation free), then 'valid' and efficient inference on θ_1 or functions thereof on the basis of

$$f(y_t | X_{t-1}, \ x_t, Y_{t-1}, \theta_1) \tag{14}$$

is possible (where 'valid' means no loss of information). This conditional model can be represented in a specific form, usually the Autoregressive Distributed Lag model is chosen for this purpose:

$$A(L)y_t = B(L)x_t + u_t. \tag{15}$$

Note again the comments on implicit and explicit assumptions mentioned with respect to the VAR representation of (12), which equally apply to this case. The final stage of modelling is to consider whether further restrictions on (15) can be imposed without loss of information. Examples are common dynamic factors, or other exclusion restrictions.

Destructive testing (for example by digesting the (mis-)specification test menus of the econometric software package PC-GIVE) aims at weeding out the invalid models. Remaining rival models are still 'comparable via the DGP' (Hendry, 1993, p. 463). Encompassing tests are supposed to serve this purpose (see chapter 7, section 4.5 below). The Wald test is the preferred tool for, among others, testing common factors. One reason is computational ease (p. 152), but the nature of the Wald test matches the general to specific modelling strategy quite well. This test is also used in the context of encompassing tests (p. 413).[9]

Hendry's justification for general to specific is based on the invalidity of the test statistics if inference commences with the simple model:

> every test is conditional on arbitrary assumptions which are to be tested *later*, and if these are rejected all earlier inferences are invalidated, whether 'reject' or 'not reject' decisions. Until the model adequately characterizes the data generation process, it seems rather pointless trying to test hypotheses of interest in economic theory. A further drawback is that the significance level of the unstructured sequence of tests actually being conducted is unknown. (p. 255)

Four objections to this argument can be made.

First, even the most general empirical model will be 'wrong' or mis-specified. Because of this possibility, Hendry (p. 257) advises testing the general model, but does not explain how to interpret the resulting test statistics. The validity of the general model is in the eyes of the beholder.[10] It is revealing that Hendry's interest has shifted from systems of equations to single equations (see p. 3), which is hard to justify from a general-to-specific perspective.

Second, one does not need an 'adequate' statistical representation of the DGP (using sufficient statistics) in order to make inferences. In many cases, stylized facts, a few figures or rough summary statistics (averages of trials in experimental economics; eyeball statistics) are able to do the work (note, for example, that statistical modelling in small particle physics is a rarity; the same applies to game theory). A crucial ingredient to a theory of reduction should be a convincing argument for the optimal level of parsimony (simplicity). No such argument is given.

Third, pre-test bias keeps haunting econometricians, whether they use stepwise regression, iterative data-mining routines, or general-to-specific modelling. Significance levels of the COMFAC test may be known asymptotically in a general-to-specific modelling strategy (as COMFAC uses a series of independent tests) (see e.g. pp. 152–3), but the sampling properties of most other test statistics *and even this one* are obscure outside the context of repeated sampling (Neyman–Pearson) or valid experimental design (Fisher).

A more general fourth objection can be raised. Would econometrics be better off if the general-to-specific methodology were to be adopted? This is doubtful. Empirical econometrics is an iterative procedure, and very few econometricians have the discipline or desire to obey the general-to-specific straitjacket. Not surprisingly, econometric practice tends to be gradual approximation, often by means of variable addition. Unlike Hendry's warnings, this practice may yield interesting and useful empirical knowledge. Consider one of the best examples in recent econometric research: the analysis of the Permanent Income Hypothesis, mentioned above in the discussion of the VAR approach. This literature, spurred on by publications of Hall, Campbell and Mankiw, Deaton, Flavin and others, is an example of fruitful variable addition. First, there was the random walk model – Harold Jeffreys would have loved this starting point. Then came excess sensitivity and excess smoothness. These problems had to be explained, and were so, by liquidity constraints among others. New variables were added to the specifications, and better (more adequate) empirical approximations were obtained (where those approximations could be directly related to advances in economic theory). This literature, not hampered by an overdose of testing and encompassing, may not carry the reductionists' praise, but the economics profession tends to disagree.[11] The empirical literature on the permanent income hypothesis is viewed as a rare success story in macro-econometrics, indeed one of the few cases where econometric analysis actually added to economic understanding of macro-economic phenomena – a case where the alternation of theory and empirical research produced a better understanding of economics.[12]

4.4 Falsificationism and the three cheers for testing

An elaboration of Fisher's approach to statistics is the rise of diagnostic testing in econometrics. Fisher claims that proper randomization justifies a normality assumption. But economists rarely randomize, and hence cannot be sure that the estimates have the desired properties. They use unique data, often time series, for which the tacit statistical assumptions are seldom warranted. Instead of randomization, one may design a model such that the statistical assumptions are not blatantly violated. If diagnostic validity tests suggest that the model is mis-specified, the modeller has to change the model.

Indeed: 'The three golden rules of econometrics are test, test and test;[13] that all three rules are broken regularly in empirical applications is fortunately easily remedied' (Hendry, [1980] 1993, pp. 27–8). Testing is proclaimed the main virtue of a scientific econometrics. This idea is related to a philosophy of science that was taught during Hendry's years at the London School of Economics by Popper and Lakatos. Their methodological falsificationism and the methodology of scientific research programmes frequently recur in Hendry's writings.

In chapter 1 it was shown that, according to Popper, one can never prove a theory right, but one may be able to falsify it by devising a 'crucial experiment'. Falsifiability separates science from metaphysics. Real scientists should formulate bold conjectures and try not to verify but to falsify. As argued in chapter 1, there are numerous problems with this view.[14] A crucial experiment is probably rare in physics (it may even be a scientific myth), and even more so in economics. Indeed, economics has hardly any experimentation to begin with (and where there is experimentation – as in game theory – econometrics is rarely invoked for purposes of inference). The claim that econometric modelling may serve as a substitute for the experimental method (uttered by many econometricians, among others Goldberger, 1964; also Morgan, 1990, pp. 9–10, 259) deserves scepticism. Finally, for methodological falsificationism, one needs bold conjectures. The empirical models of reductionist exercises are but a faint reflection thereof: they are very able representations of the data, no more, no less.[15]

So much for Popper, Hendry's favourite philosopher. His second idol is Lakatos. Lakatos (1970) noted that, in practice, few scientific theories are abandoned because of a particular falsification. Moreover, scientists may try to obtain support for their theories. He accepted a 'whiff of inductivism'. In order to appraise scientific theories, Lakatos invented the notion of 'progressive' and 'degenerative' scientific research programmes, discussed in chapter 1. A research programme is theoretically progressive if it predicts some novel, unexpected fact. It is empirically progressive if some of those predictions are confirmed. Finally, a research programme is heuristically progressive if it is able to avoid or even to reduce the dependency on auxiliary hypotheses that do not follow from the 'positive heuristic', the general guidelines of the research programme.

It is very difficult to encounter a single novel fact in Lakatos' sense in the econometric literature, and the papers of reductionism are no exception. Moreover, it is totally unclear what kind of heuristic in an economic research programme drives reductionists' empirical investigations, apart from a few rather elementary economic relations concerning consumption or money demand. Hendry, for example, does not predict novel facts. What he is able to do is provide novel interpretations of given facts, which is something entirely different. Recalling Lakatos' blunt comments on statistical data analysis in the social sciences, it will be clear that it is not justifiable to invoke Lakatos without providing the necessary ingredients of a research programme, in particular the driving 'heuristic'. Hendry does not provide a satisfactory heuristic unless elementary textbook macro-economics combined with the view that people behave according to an error-correcting scheme would qualify.[16]

However, Hendry (1993, p. 417) also applies the notion of progressiveness to his methodology itself, rather than to the economic implementations:

> The methodology itself has also progressed and gradually has been able both to explain more of what we observe to occur in empirical econometrics and to predict the general consequences of certain research strategies.

Here a 'meta research programme' is at stake: not neoclassical economics, for example, but the *method* of inference is regarded as a research programme, with an heuristic, and all the other Lakatosian notions (hard core, protective belt, novel facts). Hendry's positive heuristic is then to 'test, test and test'. The hard core might be the notion of a DGP to be modelled using 'dynamic econometrics'. Probably the most promising candidate for a 'novel fact' is the insight that, if autocorrelation in the residuals is removed by means of a Cochrane–Orcutt AR(1) transformation, this may impose an invalid common factor: the transformed residuals may behave as a white noise process but they are not innovations. This is an important insight. Whether the stated heuristic is sufficient for defining a research programme, and whether this novel fact satisfies the desire for progression, is a matter on which disagreement may (and does) exist.

Apart from the dubious relation to the philosophy of science literature, there is a very different problem with Hendry's golden rules, which is that the statistical meaning of his tests is unclear. Standard errors and other 'statistics' presented in time series econometrics do not have the same interpretation for statistical inference as they have in situations of experimental data. In econometrics, standard errors are just measurements of the precision of estimates, given the particular measurement system (econometric model) at hand. They are not test statistics. At times,

Hendry seems to agree with this interpretation. For example (Hendry, 1993, p. 420),

> Genuine *testing* can therefore only occur after the design process is complete and new evidence has accrued against which to test. Because new data has been collected since chapter 8 was published, the validity of the model could be investigated on the basis of Neyman–Pearson (1933) 'quality control' tests.

This statement also suggests that encompassing tests (discussed in the next section) are not 'genuine' tests (indeed, sometimes encompassing is presented as a form of mis-specification testing).

The reference to Neyman–Pearson in the quotation is only partly valid. Neyman–Pearson emphasize the context of decision making in repetitive situations (hence, the theory is based on repeated sampling). In Hendry's case, a number of additional observations has been collected, but the aim is not decision making but inference.

Occasionally, instead of the words 'statistical tests', the words 'diagnostic checks' are used. But not in the trinity, the three golden rules. Would they have the same aural appeal if they were 'check, check and check', or 'measure, measure and measure'? It is doubtful, although as rules they seem more appropriate. Econometricians measure rather than test, and the obsession with testing is rather deplorable. Indeed, a final problem with excessive testing is that, once the number of test statistics exceeds the number of observations used to calculate them, one may wonder how well the aim of reduction has been served, and what the meaning of the tests really is. Hendry and Ericsson (1991) provide an example. Given ninety-three annual observations, forty-six test statistics are reported. If standard errors of estimates are included, this number grows to 108. It grows further once the implicit (eyeball) test statistics conveyed by the figures are considered.

Two final remarks on statistical tests: first, there is always the problem of choosing the appropriate significance level (why does 5% deserve special attention?) – a problem not specific to Hendry's methodology; second, a statistical rejection may not be an economically meaningful rejection. If rational behaviour (in whatever sense) is statistically rejected (even if the context is literally one of repeated sampling) at some significance level, this does not mean that the objects of inference could increase their utility by changing behaviour. Money-metric test statistics would be more informative in econometric inference than statistical tests, but such tests are not discussed by Hendry or most other econometricians (Hashem Pesaran and Hal Varian are exceptions). This issue is of importance in many (proclaimed) tests of perfect markets currently presented in the finance literature.

4.5 *Model design and inference*

Hendry argues that all models are derived from the DGP and, therefore, the properties of the models are also derived entities. One does not need to invoke a DGP to arrive at this important insight of non-experimental data analysis (e.g., replace 'DGP' by 'data') – after all, what is this thing called 'DGP'? However, Hendry's emphasis on this insight deserves praise. Indeed (Hendry, 1993, p. 73),

> the consequence of given data and a given theory model is that *the error is the derived component*, and one cannot make 'separate' assumptions about its properties.

One of the strengths of Hendry's methodological observations is his recognition of the importance of model design. I agree that models are econometricians' constructs, and the residuals are '*derived* rather than *autonomous*' and, hence, models can 'be designed to satisfy pre-selected criteria' (p. 246). The distribution of the residuals cannot be stated *a priori*, as may be done in specific situations of experimental design. The 'axiom of correct specification' does not hold, which is why the econometrician has to investigate the characteristics of the derived residuals. Model design aims at constructing models that satisfy a set of desired statistical criteria. Models should be sufficient statistics.

However, this has strong implications for the interpretation of the statistical inferences that are made in econometric investigations. Most importantly, the tests are not straightforward instances of Neyman–Pearson or Fisherian tests. Accommodating the data (description) is crucially different from inference (prediction), in particular if the model that accommodates the data is not supported by *a priori* considerations.[17]

In the case of model design, the model simultaneously 'designs' its hypothetical universe. Hence, the model, given the data, cannot be used for the purpose of inference about the universe as the universe is constructed during the modelling stage, and is unique to the particular set of data which were used to construct it. Fisher (1955, p. 71), criticizing repeated sampling in the Neyman–Pearson theory, argues:

> if we possess a unique sample in Student's sense on which significance tests are to be performed, there is always, as Venn ([1866] 1876]) in particular has shown, a multiplicity of populations to each of which we can legitimately regard our sample as belonging: so that the phrase 'repeated sampling from the same population' does not enable us to determine which population is to be used to define the probability level, for no one of them has objective reality, all being products of the statistician's imagination.

This does not imply that Fisher rejected statistical inference in cases of unique samples. However, he thought experimental design a crucial element for valid inference. Economic theory is not an alternative to experimental design, in particular in macro-economics where many rival theories are available. Model design (accommodation) is not an alternative either, if the *ad hoc* element in modelling reduces the prior support of the model. The real test, therefore, remains the ability to predict out of sample and in different contexts. Friedman (1940, p. 659) made this point when he reviewed Tinbergen's work for the League of Nations:

> Tinbergen's results cannot be judged by ordinary tests of statistical significance. The reason is that the variables with which he winds up . . . have been selected after an extensive process of trial and error *because* they yield high coefficients of correlation.

This conveys the same message as Friedman's reply (in the postscript to Friedman and Schwartz, 1991) to Hendry and Ericsson (1991). Indeed, Hendry (1993, pp. 425–6) also argues that a real test is possible if (genuinely) new data become available, whereas the test statistics obtained in the stage of model design 'demonstrate the appropriateness (or otherwise) of the *design* exercise'.

The position of Friedman and Schwartz, who do not rely on modern econometric methods but are aware of the fundamental issues in statistics, might be related to Leamer's (1978) emphasis on sensitivity analysis, in combination with a strong emphasis on the importance of new data. Econometrics may not be the only, or most useful, method to find, or test, interesting hypotheses. Theory is an alternative, or even a careful (non-statistical) study of informative historical events (like the stock market crash).[18] There are few occasions where statistical tests have changed the minds of mainstream economists (with liquidity constraints in consumption behaviour arguably being a rare exception).

The hypothesis that most human behaviour follows an error-correcting pattern has not been accepted in economic textbooks, whereas Hall's consumption model and the more recent related literature is a standard topic in such books. Why is this so? Perhaps it is because the adequacy criteria in model design do not necessarily correspond to the requirements for an increase in knowledge of economic behaviour. Even the ability to 'encompass' Hall's model does not necessarily contribute to progress in economic knowledge. Encompassing, Hendry (1993, p. 440) argues,

> seems to correspond to a 'progressive research strategy' . . . in that encompassing models act like 'sufficient statistics' to summarize the pre-existing state of knowledge.[19]

As noted above, this is not what Lakatos meant by progress. Hendry is able to provide novel interpretations of existing data or models, but this is not equivalent to predicting novel facts. However, encompassing might provide a viable alternative to the Popperian concept of verisimilitude (closeness to the truth). Popper's formal definition of verisimilitude has collapsed in view of the problem of false theories. Encompassing does not necessarily suffer from this problem due to the availability of pseudo maximum likelihood estimators (where the models do not have to be correctly specified). Here is a potential subject where econometrics may contribute to philosophy of science.

4.6 Alchemy or science?

The quest for the Holy Grail resulted in a shattering of the brotherhood of the Round Table. Still, some of the knights gained worship. A quest for the DGP in econometrics is more likely to gain econometricians disrespect. Like making gold for the alchemists, searching for the DGP is a wrong aim for econometrics. Relying on the trinity of tests may be respectable, but it does not deliver the status of science. The Popperian straitjacket does not fit econometrics (or any other science). Model design makes econometrics as scientific as fashion design. Modelling is an art, and one in which Hendry excels.

Occasionally, Hendry interprets models as 'useful approximations' (1993, p. 276). Karl Pearson, a founder of statistics and an early positivist, argued along those lines, and so did other great statisticians in the tradition of British empiricism (in particular, Fisher and Jeffreys). Of course, the issue is: approximations to what? If the answer is a Holy Grail, known as the DGP, then econometrics is destined to be a branch of alchemy or, worse, metaphysics. Popperian cosmetics will not render econometrics a scientific status. If, on the other hand, the answer is approximations to the data, helping to classify the facts of economics, econometrics may join the positivist tradition which, in a number of cases, has yielded respectable pieces of knowledge. Criticism is an important ingredient of this positivist tradition, but not a methodological dogma.

5 Sensitivity analysis

5.1 Extreme Bounds Analysis

Whereas Sims aims at avoiding the problem of specification uncertainty by evading specification choices (profligate modelling without incredible restrictions), and Hendry deals with the problem by careful testing of the assumptions which are at stake, Ed Leamer provides a very different perspective. He regards, contrary to Sims, specification choices as unavoidable, and, contrary to Hendry, test statistics as deceptive. The only option which can make econometrics credible is by showing how inferences change if assumptions change. In other words, not only the final result of a 'specification search' should be reported, but the sensitivity of parameters of interest on eligible changes in the specification. An interesting economic question is rarely whether a parameter differs significantly from zero and should be deleted or not. The magnitude of a particular (set of) parameter(s) of interest, and the sensitivity to model changes, is of much more importance. Extreme Bounds Analysis (EBA) is a technical tool to obtain information on this issue. EBA itself is a technique, not a methodology. But the underlying argument, sketched above, is methodologically very different from other econometric methodologies.

Consider a regression model $y = X\beta + u$, $u \sim \mathrm{N}(0, \sigma^2\mathrm{I})$. A researcher is interested in how sensitive a subset of parameters of interest, β_i, is to changes in model specification. For this purpose, one needs to distinguish 'free' variables, X_f, from 'doubtful' variables, X_d. A parameter of interest may be free or doubtful, depending on the perspective of a researcher (see McAleer, Pagan and Volcker, 1985, who note that the fragility of estimates is sensitive to the subjective choice of classification). For the sake of clarity I will assume that the parameters of interest belong to the free variables. If a researcher fiddles with X_d (for example, by dropping some of them from the specification), it is possible that the estimates of β_i will change. This depends on the collinearity of the variables as well as the contributions of the free and doubtful variables to explaining the variation in y. EBA is an attempt to convey how sensitive β_i is to such changes in model specification (Leamer, 1978, chapter 5; Leamer, 1983).

As a first step, specification changes are limited to linear restrictions on the general model, $R\beta = r$. If the least squares estimator is denoted by $b = (X'X)^{-1}X'y$ then the constrained least squares estimator is

$$b_c = b - (X'X)^{-1}R'(R(X'X)^{-1}R')^{-1}(Rb - r). \tag{16}$$

Hence, any desired value for the constrained least squares estimates can be obtained by an appropriate choice of R and r. But even ardent data-miners tend to have a tighter straitjacket than the restrictions $R\beta = r$. Leamer (1978, p. 127), therefore, considers the special case of $r = 0$. In this case, it can be shown that the constrained least squares estimates lie on an ellipsoid; this ellipsoid defines the extreme bounds.

Extreme bounds convey two messages. The first is directly related to specification uncertainty. This message can equally well be obtained from more traditional approaches of diagnostic testing. An example is theorem 5.7 (p. 160), which proves that there can be no change in the sign of a coefficient that is more significant (i.e. has a higher *t*-statistic) than the coefficient of an omitted variable. This can be generalized to sets of regressors, as for example has been done in McAleer, Pagan and Volcker (1985, Proposition 2b), who transform EBA to a comparison of χ^2 test statistics of the significance of the set of doubtful parameters d relative to the significance of the parameters of interest i. Their condition for fragility, $\chi^2_d > \chi^2_i$, is a necessary (but not sufficient) one (for sufficiency, one also has to know the degree of collinearity between the doubtful variables and the variables of interest). It can be argued that this traditional approach is easier to carry out and even more informative (as it shows that by introducing statistically insignificant variables to the doubtful set, the 'robustness' of the parameters of interest increases). It also avoids a specific problem of EBA that a 'focus variable' in Leamer's framework becomes automatically 'fragile' once it is dubbed doubtful.[20]

A second message, though, distinguishes EBA from more traditional approaches. This message is about the numerical value of parameters of interest, and is related to importance testing (instead of significance testing). A conventional significance test, Leamer (1978) argues, is but a cumbersome way of measuring the sample size. It is of more interest to know how the value of parameters change if the model specification is changed.

Another difference with the reductionist programme of Hendry is related to the problem of collinearity. According to Hendry, collinearity is to be combated by orthogonal re-parameterization. Leamer counters that collinearity is related to the question of how to specify prior information. The fact that standard errors will be high if regressors are correlated is not necessarily a problem to Leamer.

EBA is proposed as a Bayes–non-Bayes compromise. The motivation given for EBA is Bayesian. Parameters are random and have probability distributions, but, as in Keynes' and B. O. Koopman's writings, it is argued that precise prior probability distributions are hard to formulate. Instead of specifying them explicitly, Leamer advises searching over a range of more or less convincing specifications.

5.2 Sense and sensitivity

While Hendry and other reductionists intend measuring the amount of mis-specification from the data, Leamer (1983) argues that it would be a 'remarkable bootstrap' if you could do so. In addition to the familiar standard errors reported in econometric investigations (related to the sample covariance matrix), one should consider the importance of a mis-specification covariance matrix M, which largely depends on the *a priori* plausibility of a model: 'One must decide independent of the data how good the nonexperiment is' (p. 33). According to Leamer, M figures as Lakatos' protective belt, which protects hard core propositions from falsification.

Leamer explicitly refers to the divergence between Fisherian statistics as used by agricultural experimenters and econometrics. This difference is measured by M. M may be small in randomized experiments, but it may be quite large in cases of inference based on non-experimental data. The resulting specification uncertainty remains whatever the sample size is (although, if larger samples become available, some econometricians increase the complexity of a model by, for example, using more flexible functional forms, thereby reducing the mis-specification bias).

One option to decrease M is to switch to experimental settings and randomize. This brings us into the Fisherian context of statistical research, but as noted earlier is not much practised. This option is sometimes chosen by economists (and increasingly frequently by game theorists). Another approach to reduce M is to gather qualitatively different kinds of 'nonexperiments'. Friedman's studies of consumption and money are typical examples of this approach.

The third response is to take mis-specification and data-mining for granted, but to analyse how sensitive results are to re-specifying the model. EBA is a method designed for this purpose, perhaps not the best method as the bounds are more 'extreme' than the bounds of a 'thoughtful data miner'. Many specifications that underlie points on the ellipsoid will belong to entirely implausible models. Paraphrasing Jeffreys ([1939] 1961, p. 255), whoever persists in looking for evidence against a hypothesis will always find it. Leamer's method tends to exaggerate the amount of scepticism (although papers in the reductionist programme that do not investigate sensitivity at all, exaggerate the amount of empirical knowledge that can be obtained from econometric studies).

Granger and Uhlig (1990) suggest making the bounds more reasonable by not taking the extremes, and by imposing constraints in addition to $Rb = r$. In particular, they propose considering only those specifications that have an R^2 of at least 90 or 95% of the maximum R^2 that can be obtained. This proposal is very *ad hoc*. It does not remove the real problem of EBA: whatever the bound, it is not clear whether the resulting specification is plausible. Plausibility is hard to formalize, but the minimum description length (MDL) principle, discussed in chapter 5, may serve as a heuristic guide. Goodness of fit (more precisely, the likelihood) is one element of that principle. However, relying on R^2 as a statistic for goodness of fit is not to be recommended, and even $\bar{R}^2$ is not a very good guide for a sensible sensitivity analysis, as those statistics are bad guides in small samples (Cramer, 1987, shows that R^2 in particular is strongly biased upwards in small samples, and both R^2 and $\bar{R}^2$ have a large dispersion; Cramer advises not using these statistics for samples of less than fifty observations).

A sensible refinement of extreme bounds sensitivity analysis can be found in Levine and Renelt (1992). They discuss the fragility of cross-country growth regressions. Among other things, they consider regressions of GDP *per capita*, y $(n \times 1)$, on a set of free variables X_f $(n \times k)$, including investment shares, initial level of GDP, education and population growth. For calculating extreme bounds, one of those variables in turn is set apart as the variable of interest, x_i $(n \times 1)$. In addition, up to three doubtful variables, X_d $(n \times l,\ l \leq 3)$ that may potentially be of importance in explaining cross-country growth differences are included. These variables are taken from a list obtained from analysing previous empirical studies.

The result is linear equations of the following kind:

$$y = \beta_i x_i + X_f \beta_f + X_d \beta_d + u. \tag{17}$$

The differences from Leamer's original approach are:

- the restriction $1 \leq 3$
- the (potentially very large) list of doubtful variables is restricted to seven indicators that are argued to be a 'reasonable conditioning set'
- a further selection is used: for every variable of interest x_i, the pool of variables from which to choose X_i excludes those variables that are thought on *a priori* grounds to measure the same phenomenon as x_i.

The extra restrictions lead to smaller but more convincing extreme bounds. This makes sense, as the bounds in Leamer's original approach may be very wide, resulting from nonsense specifications.

Levine and Renelt not only report the extreme bounds, but also the set of variables that yields those bounds. This gives further information to readers who can judge for themselves whether this particular specification makes sense at all. The authors also consider the effect of data quality on the bounds. After an impressive data analysis (without statistical validity testing, though) they conclude that the correlation between output and investment growth is 'robust', but a large assortment of other economic variables are not robustly correlated with growth: 'the cross country statistical relationship between long-run average growth and almost every particular macroeconomic indicator is fragile' (Levine and Renelt, 1992, p. 960). This study has become a classic in the empirical growth literature. With qualifications, EBA is a very useful method. More importantly, the argument underlying EBA deserves more consideration in econometrics.

6 Calibration

6.1 Elements of calibration

Kydland and Prescott (1994, p. 1) argue that performing experiments with actual people at the level of national economics is 'obviously not practical, but constructing a model economy inhabited by people and computing their economic behavior is'. Their 'computational experiment' re-introduces the old view that econometric models are an alternative to physical experiments.[21] A model economy is calibrated 'so that it mimics the world along a carefully specified set of dimensions'. Those dimensions are usually first and second moments of some business cycle phenomena and compared with the actual moments. The resulting model enables counterfactual analysis which can be used for policy evaluation. The choice of the model (and the dimensions on which it is evaluated) depends entirely on the particular question of interest.

Calibration consists of a number of elements. The most important is a 'well-tested theory'. Kydland and Prescott's definition of *theory* (p. 2) is taken from Webster's dictionary, i.e. 'a formulation of apparent relationships or underlying principles of certain observed phenomena which has been verified to some degree'. As an example of this (a-Popperian) definition, they mention neoclassical growth theory which 'certainly' satisfies the criterion in the definition. One of the problems of econom(etr)ics is that not everyone agrees. To make things worse, no operational definition of verification is presented. The obvious candidate based on statistical criteria is explicitly rejected: calibration is presented as an alternative to econometrics (in the modern, narrow, sense of statistical analysis of economic data).

The second important element of a calibrated model is a set of parameter values. Unlike different approaches in econometrics, calibrators do not estimate these parameters in the context of their own model. Instead, they pick estimates found elsewhere, often in (micro-)econometric investigations related to specific features of the model to be calibrated. A great advantage of the calibration approach is the utilization of 'established' findings, including salient 'stylized facts' of economies, which are imputed into the model. Examples are labour- and capital shares for the calibration of (Cobb–Douglas) production functions, or a unitary elasticity of substitution between consumption and leisure.

The third element of calibration is the actual process of calibrating. The 'well-established theory' plus a selection of 'deep parameters' are not sufficient to specify the dynamics of the general equilibrium model: 'Unlike theory in the physical sciences, theory in economics does not provide a law of motion' (Kydland and Prescott, 1994, p. 2). This law is computed by computer simulations in such a way that the 'carefully specified set of dimensions' of the model are in accordance with reality. The metaphor of a thermometer is invoked to justify this step. The level of mercury in water at freezing point and boiling water is indicated and it is known that this corresponds with 0° and 100° Celsius, respectively. Add the information that mercury expands approximately linearly within this range and the measurement device can be calibrated. Similarly, a model economy should give approximately correct answers to some specific questions where the outcomes are known. In practice, simulated moments are compared with actual moments of business cycle phenomena. If correspondence is acceptable, then it is argued that a necessary condition is satisfied for having confidence in answers to questions where the 'true' answer is not known *a priori*.

6.2 The quest for deep parameters

The most important characteristic of calibration is its strong reliance on economic theory. According to Eichenbaum (1995, p. 1610),

> the models we implement and our empirical knowledge have to be organised around objects that are interpretable to other economists – objects like the parameters governing agents' tastes, information sets and constraints. To the extend that econometricians organise their efforts around the analysis of the parameters of reduced form systems, their output is likely to be ignored by most of their (macro) colleagues.

A similar argument was made in chapter 5 on simplicity: structural relations are more informative than idiosyncratic reduced form equations. It is also the argument of the Cowles Commission, and I think it is a valid explanation for the limited success of VAR business cycle research. The difference between calibrators and Cowles econometricians is that calibrators reject Haavelmo's probability foundations in econometrics, i.e. formal Neyman–Pearson methods of inference.

One may wonder how much calibration adds to the knowledge of economic structures and the deep parameters involved. Calibration is an interesting methodological procedure, as it extends the merits of empirical findings to new contexts. Micro estimates are imputed in general equilibrium models which are confronted with new data, not used for the construction of the imputed parameters. Imputation yields over-identifying restrictions which are needed to establish the dynamic components ('laws of motion') of the model. However, this procedure to impute parameter values into calibrated models has serious weaknesses and overestimates the growth of knowledge about structures and deep parameters.

First, few 'deep parameters' (those related to taste, technology, information and reactiveness) have been established at all, by econometric or other means (Summers, 1991). Like Summers, Hansen and Heckman (1996, p. 90) are sceptical about the knowledge of deep parameters: 'There is no filing cabinet full of robust micro estimates ready to use in calibrating dynamic stochastic general equilibrium models.'

Second, even where estimates are available from micro-econometric investigations, they cannot be automatically imported (as is done in calibration exercises) into aggregated general equilibrium models: usually, they are based on partial equilibrium assumptions. Parameters are context dependent. Calibrators suffer from a realist illusion if they think that estimates can be transmitted to other contexts without problem.[22] Hansen and Heckman (1996, p. 90) claim that 'Kydland and Prescott's account of the availability and value of micro estimates for macro models is dramatically overstated'. Kydland and Prescott combine scientific realism about 'deep parameters' with the opposite view about the aggregate equilibrium models. This micro-realism versus macro-instrumentalism is an unusual methodological stance.

Third, calibration hardly contributes to growth of knowledge about 'deep parameters'. These deep parameters are confronted with a novel context (aggregate time-series), but this is not used for inference on their behalf. Rather, the new context is used to fit the model to presumed 'laws of motion' of the economy. The characteristics of the model (distributions or statistics of parameters of interest) are tuned to the known characteristics of the data, like a thermometer which is tuned to the (known) distribution of temperatures. The newly found parameters for the laws of motion are not of primary interest (arguably, they are not 'deep' but idiosyncratic, i.e. specific, to the model which is at stake) and, therefore, not subject to testing. This is not necessarily bad scientific practice: the merits of testing are highly overvalued in mainstream econometrics and the calibrators' critique of formal hypothesis testing is not without grounds. But without criteria for evaluating the performance of a thermometer, its reported temperature is suspect. Statistics is a useful tool for measurement and Kydland and Prescott do not give a single argument to invalidate this claim (Eichenbaum, 1995, is more optimistic about the possibility of using probabilistic inference in combination with calibration, as he provides some suggestions to introduce Bayesian inference and the Generalised Method of Moments procedure into the calibration methodology).

This leads to the fourth weakness. The combination of different pieces of evidence is laudable, but it can be done with statistical methods as well (Bayesian or non-Bayesian approaches to pooling information). This statistical approach has the advantage that it takes parameter uncertainty into account: even if uncontroversial 'deep parameters' were available, they would have standard errors. Specification uncertainty makes things even worse. Neglecting this leads to self-deception.

6.3 Calibration and probabilistic inference

Probability plays a limited role in calibration where stochastic uncertainty in a competitive equilibrium model results in a stochastic process which is simulated by a computer. The 'shocks' to the model (e.g. in technology) have a well-defined probability distribution (imposed by the researcher). Sampling theory is invoked to measure the distribution of statistics of interest generated by the model. The computer can generate as many (stochastic) repetitions as desired, and the notions of repeated sampling or collective is justified in this context.[23] It should be noted that the relation between the models and reality is a totally different issue to which this use of sampling theory does not relate.

As argued above, criteria for evaluating the results of a calibration exercise are not given. Calibrators do not rely on a model in which goodness of fit and other statistical arguments are decisive criteria: 'the goal of a computational experiment is not to try to match correlations' (Kydland and Prescott, 1994, p. 8). Statistical theory is not invoked to validate economic models, it is 'not a useful tool for testing theory' (p. 23). Curve fitting (e.g. by means of the Hodrick–Prescott filter), based on least squares techniques, is not supplemented with probabilistic arguments. One of the invoked arguments against significance testing is, that all models are 'false', and any significance test with a given size will ultimately reject the model, if a sufficient number of observations is obtained. For example, Eichenbaum (1995, p. 1609) argues,

> Since all models are wrong along some dimension of the data the classic Haavelmo programme is not going to be useful in this context [Probability approach in econometrics, 1944]. We do not need high-powered econometrics to tell us that models are false. We know that.

The calibration approach goes a step further than the Cowles Commission in its trust in theory, while the probability approach to econometrics is abandoned as 'utopian' (p. 1620). But total rejection of statistical analysis may lead to unnecessary loss of information.

Kydland and Prescott (1994, p. 23) claim that a useful way to measure empirical adequacy is to see 'whether its model economy mimics certain aspects of reality'. No arguments are given as to how to assess the quality of the correspondence and how to select the list of relevant aspects (moment implications of the model). This is a serious weakness in the calibration approach. How trustworthy is a model with a 'significant' empirical defect in some of its dimensions, even if it is not this particular dimension which is the one of theoretical interest?

Pagan (1995) suggests that calibrated models are invalid reductions of the data. Kydland and Prescott deny that this is an interesting criticism. For example, they argue that

> all models are necessarily abstractions. A model environment must be selected based on the question being addressed... The features of a given model may be appropriate for some questions (or class of questions) but not for others. (1996, p. 73)

There is no unique way to compare models and rank them, as this is a context-dependent activity. The authors invoke Kuhn (1962, pp. 145–6), who argues that the resolution of scientific revolutions will not be settled instantly by a statistical test of the old versus the new 'paradigm'. Kydland and Prescott (1996, p. 73, n. 4) again quote Kuhn (1962, p. 145) who argues: 'Few philosophers of science still seek absolute criteria for the verification of scientific theories.' They do not quote the remainder of Kuhn's argument, which says '[n]oting that no theory can ever be exposed to all possible relevant tests, they ask not whether a theory has been verified but rather about its probability in the light of the evidence that actually exists'. Kuhn continues to show the difficulties due to the fact that probabilistic verification theories all depend on a neutral observation language. Kuhn's incommensurability theory, which denies such a neutral language, undermines probabilistic testing of rival theories. As a result, he concludes, 'probabilistic theories disguise the verification situation as much as they illuminate it' (p. 146). This phrase is cited with approval by Kydland and Prescott.

However, Kuhn's argument differs fundamentally from the calibrators'. The latter pretend that a neutral language does exist and even yields information about 'true' deep parameters. Where commensurability prevails, Kuhn (p. 147) holds that it 'makes a great deal of sense to ask which of two actual and competing theories fits the facts *better*'. A different motivation for rejecting probabilistic inference should be presented. Kuhn is the wrong ally.

An effort to provide a motivation is made by the claim that some model outcomes should obey known facts, but 'searching within some parametric class of economies for the one that best fits a set of aggregate time series makes little sense' (Kydland and Prescott, 1994, p. 6). The best fitting model may not be the most informative on the question at stake. Moreover, it may violate 'established theory' (p. 18), which should not be readily forsaken. One of Prescott's most revealing titles of a paper is 'Theory ahead of business cycle measurement' (1986).

The definition of 'theory' inspires the rejection of statistical inference. Following Lucas (1980), theory is conceived as a set of instructions for building an imitation economy to address specific questions. It is 'not a collection of assertions about the behaviour of the actual economy. Consequently, statistical hypothesis testing, which is designed to test assertions about actual systems, is not an appropriate tool for testing economic theory' (Kydland and Prescott, 1996, p. 83). This negation in the definition of 'theory' is curious. Few of the great theorists in economics and the sciences would accept it. But it is true that *testing* theory is a rare activity and moreover, that statistical hypothesis testing is at best of minor importance for this. Eichenbaum (1995, p. 1620) is correct in his assertion that the Cowles programme, 'with its focus on testing whether models are "true", means abandoning econometrics' role in the inductive process'. This does not imply the Kydland–Prescott thesis, that probabilistic inference is not useful for empirical economics. The latter thesis is untenable.

7 Conclusion

Specification uncertainty is the scourge of econometrics: it has proven to be the fatal stumbling block for the Cowles programme in econometrics. This chapter has discussed various ways of dispelling this scourge: VAR, reductionism, sensitivity analysis, and calibration. It is useful to summarize their characteristics briefly in terms of the underlying philosophy, the probability theory, and the views on testing and simplicity. This is done in table 1.

Table 1.

	VAR	Reductionism	Sensitivity analysis	Calibration
Philosophy of science	instrumentalism	realism methodological falsification	instrumentalism	micro-realism, macro-instrumentalism
Probability theory	first frequentist, later Bayesian	Fisher, Neyman–Pearson	Bayesian	probability-aversion
Statistical testing	only relevant for determining lag length	makes econometrics scientific	not informative about substantive issues	not informative about substantive issues
Simplicity	theory imputes incredible restrictions – should be evaluated by information criterion	general to specific 'encompassing' model, should be evaluated by significance test	belongs to realm of metaphysics, but desirable feature of models	desirable feature of models

The great drawback of the VAR approach is that it produces idiosyncratic models, where the criticism of 'arbitrary exclusion restrictions' leads to a violation of the simplicity postulate. A better alternative to evaluate the effect of restrictions is Leamer's sensitivity analysis. His method may be imperfect but the idea is sound and deserves much more support than it has received. Hendry's reductionism collapses under its Popperian and probabilistic pretences. Significance testing does not make econometrics a science and the methodology is likely once again to yield idiosyncratic models: good representations of specific sets of data, but hardly successful in deriving novel knowledge on generic features of the economy. Calibration, finally, throws away the baby with the bathwater. I support the claim that all models are wrong, but there are alternative statistical approaches that are able to accommodate this claim. If econometrics were to escape the straitjacket of the Neyman–Pearson approach, calibration might return to a fruitful collaboration with econometrics.

Statistical inference may be an informative research tool but without investigating the credibility of the assumptions of such a model, and the sensitivity of the parameters of interest to changes in assumptions, econometrics will not fulfil its task. Sensitivity analysis, in its different forms (Leamer's, Friedman's) should be a key activity of all applied econometricians. This would also re-direct the research interest, away from Neyman–Pearson probabilistic tests and Popperian pretence, back to the substantive analysis of economic questions.

Notes

1. Alternative identifying information may come from linear restrictions in the model, or from restrictions on the stochastic characteristics of the model.
2. Summers (1991, p. 135) is 'Popperian' in claiming that science 'proceeds by falsifying theories and constructing better ones'. In reality, the falsification of theories is a rare event and not the main concern of scientists. In the remainder of his article, Summers is much closer to the instrumentalistic approach that is supported in this book.
3. Sims (1980) can be regarded as an extension of Liu's argument.
4. The discussion of reductionism draws heavily on Keuzenkamp (1995).
5. These authors define a Bayesian experiment as 'a unique probability measure on the product of the parameter space and the sample space' (p. 25). Upon asking the second author why the word 'experiment' had been chosen, the answer was that he did not know.
6. This conflation is explicitly recognized by Hendry (1993, p. 77). See also: 'a correct model should be capable of predicting the residual variance of an

incorrect model and any failure to do this demonstrates that the first model is not the DGP' (p. 86).

7. $P(w|y) = P(w, y)/P(y)$.
8. Hendry does not have a strong interest in non-parametric or non-linear models. This is hard to justify in a general-to-specific approach. It rules out economic models with multiple equilibria.
9. A problem with the Wald test is that it is not invariant to mathematically identical formulations of non-linear restrictions (Godfrey, 1989, p. 65). It is usually assumed that this does not affect linear restrictions, such as $\beta_1 = \beta_2$. However, an identical formulation of this striction is $\beta_1/\beta_2 = 1$, which may lead to the same problem of invariance. Hence, using a Wald-type general-to-specific test may not always be advisable. Hendry is open minded with respect to the choice of test statistics: usually there are chosen on pragmatic grounds.
10. For example, the practical starting point of most of Hendry's modelling, the autoregressive distributive lag model (in rational form $y_t = \{b(L)/a(L)\} x_t + \{1/a(L)\}u_t$), is less general than a transfer function model, $y_t = \{b(L)/a(L)\}x_t + \{m(L)/r(L)\}u_t$. This arbitrary specific starting point may lead to invalid inferences, for example, in investigating the permanent income hypothesis (see Pollock, 1992).
11. Of course, one may object that economists are not the best judges on the fruits of econometrics. This would violate a basic rule of economics (and political science): the consumer (voter) is always right.
12. Hendry's contribution to this literature is contained as chapter 10 in Hendry (1993). It is shown that an error correction model is able to 'encompass' Hall's random walk model of consumption in the context of UK data. There is no doubt that Hendry's specification provides a better approximation to the data. However, it was not Hall's point to obtain the best possible approximation. A similar remark applies to the discussion between Hendry and Ericsson (1991) and Friedman and Schwartz (1991). See also the discussion of calibration in section 5 below.
13. A footnote is added here: 'Notwithstanding the difficulties involved in calculating and controlling type I and II errors' (p. 28).
14. See e.g. the excellent treatment in Hacking (1983).
15. Darnell and Evans (1990), who make an effort to uphold the Popperian approach to econometrics (meanwhile entertaining a quasi-Bayesian interpretation of probability) similarly complain that Hendry does not deliver the Popperian goods.
16. Note, moreover, that the recent shift to co-integration, which can also be found in Hendry's writings, weakens the error-correcting view of the world. The links between co-integration and error correction are less strong than at first sight may seem (see Pagan, 1995).
17. See Howson (1988b) for a useful view on accommodation and prediction, from a Bayesian perspective. Howson claims that the predictionist thesis is false. This thesis is that accommodation does not yield inductive support to a hypothesis, whereas independent prediction does. Howson's claim is valid

(roughly speaking) if the hypothesis has independent *a priori* appeal. Dharmapala and McAleer (1995), who deal with (objective) truth values instead of degrees of belief, share Howson's claim, but ignore the condition of *a priori* support. In the kind of econometric model design, where *ad hoc* considerations play an important role and *a priori* support is weak, the predictionist thesis remains valid.

18. See also Summers (1991).
19. In the section omitted from the citation, there is a reference to Lakatos amongst others. Hendry and Ericsson (1991, p. 22) make nearly the same claim, using a very subtle change of words. Encompassing is 'consistent with the concept of a progressive research strategy... since an encompassing model is a "sufficient representative" of previous empirical findings'. One wonders whether the move from 'statistic' to 'representative' has any deeper meaning (as in the distinction between a 'test' and a 'check').
20. See also Ehrlich and Liu (1999) for a wide range of examples showing the misleading information which may result from thoughtless application of EBA.
21. The computational experiment is the digital heir to Weber's *Gedankenexperiment*. Like Weber's, it is not an experiment but an argumentation or simulation. As shown before, 'experiment' is a much abused term in econometrics; calibration follows this tradition.
22. Philosophical realism is discussed in chapter 9, section 2.
23. An additional source of uncertainty may be introduced for the parameters of the model. Bayesian methods can be used to specify such models.

8 In search of homogeneity

> Many writers have held the utility analysis to be an integral and important part of economic theory. Some have even sought to employ its applicability as a test criterion by which economics might be separated from the other social sciences. Nevertheless, I wonder how much economic theory would be changed if either of the two conditions above [homogeneity and symmetry] were found to be empirically untrue. I suspect, very little.
>
> Paul A. Samuelson ([1947] 1983, p. 117)[1]

1 Introduction

After discussing general problems of statistical and econometric inference, I will turn to a case study.[2] This chapter deals with consumer behaviour which belongs to the best researched areas in both economic theory and econometrics.

The theory of consumer behaviour implies a number of properties of demand functions. Such demand functions have been estimated and the predicted properties have been compared with the empirical results. Among the most interesting properties is the homogeneity condition, a widely accepted piece of economic wisdom which, however, does not seem to stand up against empirical tests.

This chapter discusses why the rejections were not interpreted as Popperian falsifications of demand theory. Furthermore, the role of auxiliary assumptions is discussed. Finally, the merits of Leamer's (1978) 'specification searches' for interpreting the literature on testing homogeneity are assessed.

2 The status of the homogeneity condition: theory

2.1 Introduction

The condition of homogeneity of degree zero in prices and income of Marshallian demand functions belongs to the core of micro-economic wisdom. Simply put, this condition says that if all prices and income change proportionally, expenditure in real terms will remain unchanged. This has a strong intuitive appeal. Some economists call it the absence of money illusion, they claim that it is a direct consequence of rational behaviour.

The homogeneity condition results from an idealizing hypothesis, i.e. linearity of the budget constraint which, jointly with a set of idealizing hypotheses related to the preferences of individuals, implies the *law of demand*. This law states that compensated demand curves do not slope upward. Another ideal condition that can be derived from the idealizing hypotheses is symmetry and negative definiteness of the so-called Slutsky matrix of compensated price responses. Economists have tested these conditions, and, in many cases, they have had to reject them statistically.

2.2 Falsifiable properties of demand

2.2.1 Axioms and restrictions

The economic theory of consumer behaviour is founded on a set of axioms and assumptions. Let preferences of some individual for consumption bundles q be denoted by the relations $\succeq$ ('is weakly preferred to') and $\succ$ ('is strictly preferred to'). The following requirements or 'axioms of choice' are imposed (see the excellent introduction to demand theory by Deaton and Muellbauer, 1980b, pp. 26–9; much of the following is explained in more detail in this or other intermediate textbooks on micro-economics).

Axioms of choice

(A1) Irreflexivity: for no consumption bundle q_i, it is true that $q_i \succ q_i$

(A2) Completeness: for any two bundles in the choice set, either $q_i \succeq q_2$ or $q_2 \succeq q_1$ or both

(A3) Transitivity (or consistency): if $q_1 \succ q_2$ and $q_2 \succ q_3$, then $q_1 \succ q_3$

(A4) Continuity

(A5) Nonsatiation

(A6) Convexity of preferences.

The first three axioms define what economists regard as rational behaviour. (A1) to (A4) imply the existence of a well-behaved utility function,

$$U(q_1, q_2, \dots q_n), \qquad (1)$$

where q_i denotes the (positive) quantity of commodity i ($i = 1, \dots, n$). The utility function is continuous, monotone increasing and strictly quasi-concave in its arguments, the quantities q_i. The consumer has to pay (a positive) price p_i per unit of commodity i.

The fourth axiom is not required for rational behaviour (it rules out lexicographic preferences even though these may be the taste of a perfectly rational individual) but is generally imposed for (mathematical) convenience. (A5) ensures that the solution is on the budget constraint instead of being inside the constraint (hence, there is no bliss point). (A6) allows the use of standard optimization theory but is non-essential for neoclassical demand theory.

Utility is maximized subject to the budget constraint. An important additional restriction (R) usually imposed on the consumer choice problem is linearity of this constraint:

$$\text{(R1)} \quad \sum_{i=1}^{n} p_i q_i = m, \tag{2}$$

where m is the budget (or means) available for consumption.

Yet another restriction, needed if one wants to carry out empirical analysis, is to impose a particular functional form on the utility function or derived functions (in particular the indirect utility function, and the expenditure function), and to select an appropriate set of arguments that appear in the utility function. I will summarize these constraints in:

(R2) Choice of a particular functional form.

(R3) Choice of arguments.

Are these axioms and assumptions like Nowak's (1980) 'idealizing hypotheses'? Such idealizing hypotheses are false, in the sense that they are in conflict with reality. Most economists would agree that this is the case for (A1)–(A6) and (R1)–(R3). Economic behaviour can be explained by this set of restrictions and axioms, but it is acknowledged that economic subjects do not always fit the straitjacket. This provides the food for psychologists.

Indeed, tests of the axioms (such as transitivity, and in particular tests in a situation of choice under uncertainty) have repeatedly shown that the axioms are violated by mortals – even by Nobel laureates in economics. The papers in Hogarth and Reder (1986) provide ample evidence against perfect rationality as defined above.

Not only the axioms are 'false', but the additional restrictions are also defective. We know, for example, that occasionally prices do depend on the quantity bought by a particular consumer, either because the consumer is a monopsonist, or because of quantity discounts. This would imply a non-linear budget constraint. However, such non-linearities are thought to be fairly unimportant in the case of choosing a consumption bundle.

Hence, the axioms and constraints are 'false', but still they are very useful for deriving economic theorems. One such theorem is the law of demand, to be specified below. But the axioms and further restrictions are not just instrumental in deriving ideal laws, they also serve to interpret the real world. It has even been argued that a researcher who obtains empirical findings or correlations that imply violations of any of (A1)–(A6) and (R1), is not entitled to call such correlations 'demand functions': the estimates of the parameters are not valid estimates of elasticities (Phlips, 1974, p. 34). Such violations can be found by analysing the properties of demand functions that can be deduced from the axioms and restrictions. I will now turn to this issue.

2.2.2 Demand functions and their properties

Reconsider the utility function which does not have income and prices as arguments. Furthermore, the function presented is a static one, independent of time. Both assumptions are examples of (R3). The individual agent maximizes utility subject to the budget constraint. Solving the optimization problem results in the Marshallian demand equations that relate desired quantities to be consumed to income and prices, of which the one for good i reads:

$$MD \quad q_i = f_i(m, p_1, p_2, \ldots, p_n). \tag{3}$$

It is by way of (R1) that the prices and the budget enter the demand equation.

Alternatively, it is possible to formulate demand functions in terms of a given level of utility, the so-called Hicksian or compensated demand functions:

$$HD \quad q_i = h_i(U, p_1, p_2, \ldots, p_n). \tag{4}$$

It is now possible to be more explicit about the properties of demand functions. There are four properties of interest (Deaton and Muellbauer, 1980b, pp. 43–4):

(P1) Adding up: the total value of demands for different goods equals expenditure

(P2) Homogeneity: the Marshallian demand functions are homogeneous of degree zero in prices and expenditure, the Hicksian demand functions are homogeneous of degree zero in prices

(P3) Symmetry: the cross price derivatives of Hicksian demand functions are symmetric

(P4) Negativity: the matrix of cross price derivatives of Hicksian demand functions is negative semi-definite.

It is (P4) in isolation which is often called the 'law of demand'. It is a typical example of an 'ideal law'. It implies downward sloping compensated demand functions. Occasionally, the four properties taken together are interpreted as defining demand functions (Phlips, 1974). Note that Hicks ([1939] 1946, pp. 26, 32) shows that the downward sloping compensated demand curve is (using Nowak's terminology) a concretization of Marshall's original demand function. Marshall derived the condition that the (uncompensated) demand function is downward sloping by imposing that the marginal utility of money is constant. When this idealizing assumption is dropped, the remaining results are still of interest and, one might add, closer to reality (truth).

The homogeneity property, (P2), states that if m and all p_i are multiplied by the same factor, the quantity demanded of good i (or any other good) is not changed. Hence,

$$q_i = f_i(m, p_1, p_2, \ldots, p_n) = f_i(\alpha m, \alpha p_1, \alpha p_2, \ldots, \alpha p_n), \tag{5}$$

or $\partial q_i / \partial \alpha = 0$: Nowak's (1980, p. 28) condition to qualify as an idealizing assumption. The price and income terms enter the demand equation via the budget constraint, not the utility function. Multiplying both sides of the constraint by α has no consequences for the composition of the set of quantities demanded to satisfy the budget constraint. Hence, the homogeneity condition is just an implication of the (linear) budget constraint and only a very weak condition of rational choice. Deaton (1974, p. 362) argues that it is hard to imagine any demand theory without the homogeneity condition and even identifies non-homogeneity with irrationality (in which case either (P2) or (R1) should be added to (A1)–(A3) in defining rational behaviour). Deaton is not alone in viewing (P2) as an indispensable condition. For example, Phlips (1974, p. 34) writes:

> Every demand equation must be homogeneous of degree zero in income and prices... In applied work, only those mathematical functions which have this property can be candidates for qualification as demand functions.

I will now turn to the question of whether empirical work on estimating demand functions served a particular role in 'concretizing' the theory of consumer behaviour. In particular, I will focus on testing (P2), the homogeneity condition.

For expository reasons, it is useful to write the Marshallian demand (3) in double logarithmic form and to impose linearity:

$$\ln q_i = \beta_i + \eta_i \ln m + \sum_{j=1}^{n} \epsilon_{ij} \ln p_j \qquad i, j = 1, 2, \ldots, n, \tag{6}$$

where η_i is the income elasticity of demand and ϵ_{ij} are the price elasticities. If multiplication of m and all p_i by the same factor leaves q_i unchanged (see (5)), then the following condition should hold:

$$\eta_i + \sum_{j=1}^{n} \epsilon_{ij} = 0. \tag{7}$$

This equation has been the basis for a number of empirical tests of the homogeneity condition. An alternative formulation has been used as well. This will be derived now. The price elasticity can be decomposed as:

$$\epsilon_{ij} = \epsilon_{ij}^{*} - \eta_i w_j, \tag{8}$$

where $w_j = p_j = q_j/m$ defines the expenditure share of good j. Now (6) can be rewritten as:

$$\ln q_i = \beta_i + \eta_i \left(\ln m - \sum_{j=1}^{n} w_j \ln p_j \right) + \sum_{j=1}^{n} \epsilon_{ij}^{*} \ln p_j. \tag{9}$$

As is obvious from (9) the $\eta_i w_j$ component of ϵ_{ij} is the part of the price reaction which can be neutralized by an appropriate change in m. Thus the ϵ_{ij}^{*} are known as the compensated (or Slutsky) price elasticities as opposed to the uncompensated ones. It follows from the property that $\Sigma_{j=1}^{w} = 1$ that the homogeneity condition, (7) is equivalent to:

$$\sum_{j=1}^{n} \epsilon_{ij}^{*} = 0. \tag{10}$$

Both (7) and (10) have been used to test homogeneity.

2.2.3 Homogeneity and the systems approach

In principle, there are as many demand equations as there are commodities and they should all satisfy the homogeneity condition. One single contradiction invalidates it as a general property. A test of homogeneity should, therefore, involve all equations. A natural approach would be a formal test of the homogeneity condition for all equations at the same time.

The systems approach to demand analysis supplies an opportunity for such a type of test. The empirical application of the Linear Expenditure System (LES) by Stone (1954b) paved the way for this approach.

Normally, the same functional form is used for all equations of the system. From the point of view of testing homogeneity, this is not really needed. Johansen (1981) views it as a straitjacket, reducing the empirical validity of the system. Departing from this practice, however, makes the use (or test) of other properties of demand functions, like symmetry, cumbersome. It would reduce the attraction of the systems approach as a relatively transparent, internally consistent and empirically applicable instrument of demand analysis.

Let us turn to the famous early example of an application of the systems approach: the LES. The utility function is specified as:

$$\sum_{i=1}^{n} \beta_i \ln(q_i - \gamma_i), \qquad \sum_{i=1}^{n} \beta_i = 1, \tag{11}$$

with β_i and γ_i ($\gamma_i < q_i$, $\beta_i > 0$), and $i = 1, \ldots, n$. This utility function is maximized subject to the budget equation (R1). The first order conditions for a maximum are solved for the optimizing quantities to yield the demand equations:

$$q_i = \gamma_i + \beta_i \left(m - \sum_{j=1}^{n} \gamma_i p_i \right) / p_i. \tag{12}$$

Multiplying through by p_i yields an equation for optimal expenditures as a linear function of m and the prices. Note that all prices appear in each demand equation. Note also from (12) that homogeneity is an integral part of the specification. Indeed, the LES was not invented to test demand theory, but to provide a framework for the measurement of allocation. Significance tests reported should be interpreted as referring to the accuracy of measurements (estimates), not the accuracy of the theory.

Other types of specification, however, do allow for testing the homogeneity condition. One important type makes use of duality and starts off from the indirect utility function, expressing the maximum obtainable utility for a given set (m,$\mathbf{p}$) where $\mathbf{p}$ indicates a vector of prices:

$$U(f_1(m, \mathbf{p}), f_2(m, \mathbf{p}), \ldots, f_n(m, \mathbf{p}) = V(m, \mathbf{p}). \tag{13}$$

Such a function is theoretically homogeneous of degree zero in (m, $\mathbf{p}$), but one does not need to impose this property on the specification. Using Roy's identity, one can obtain Marshallian demand functions:

$$q_i = -\frac{\partial V/\partial \mathbf{p}}{\partial V/\partial m}, \tag{14}$$

with the same functional form for all i but not necessarily homogeneous of degree zero in $(m, \mathbf{p})$. One can replace the $f_i(m, \mathbf{p})$ in (13) by $f_i(\pi)$ where π is the vector of prices $\mathbf{p}$ divided by m. Then $V(m, \mathbf{p})$ specializes to $V^*(\pi)$. Since π is invariant for the same proportional change in m and the $\mathbf{p}$, this reformulation conserves the homogeneity condition. The counterpart of (14) is then:

$$q_i = -\frac{\partial V^*/\partial \mathbf{p}_i}{\partial V^*/\partial m}. \tag{15}$$

It is essential that (15) is distinguishable from (14), for example, in the form of restrictions on the parameters to be estimated. One can compare the empirical performance of (14) with that of (15) for all equations of the system together. If that of (14) is significantly better than that of (15) one has to reject the homogeneity condition. The Generalized Leontief System of Diewert (1974) and the Indirect Translog System of Christensen, Jorgenson and Lau (1975) are examples of (14). These systems were, however, not used to test homogeneity separately from the other restrictions.

A related approach stipulates the functional form of the expenditure function. This function expresses the minimum m needed to reach a certain utility level for a given $\mathbf{p}$. Like the indirect utility function, it is a concept that reflects the utility optimizing behaviour of the consumer.

The expenditure function can be obtained by solving m in the indirect function (13) in terms of U and $\mathbf{p}$, yielding, say, $\tilde{m}(U, \mathbf{p})$. According to Shephard's lemma the partial derivatives of the expenditure function with respect to p_i are the Hicksian (or compensated) demand functions in terms of U and $\mathbf{p}$. Replacing U by the right-hand side of (13) gives the desired demand functions. Homogeneity implies that $\tilde{m}(U, \mathbf{p})$ is homogeneous of degree one in the prices: if prices double, expenditure must double as well in order to reach the same utility level. If that is indeed the case and $V^*(\pi)$ is used to replace U in the Hicksian demand functions, they are homogeneous of degree zero. However, one does not need to impose the restriction right away. By comparing the restricted and unrestricted equations, the restriction may be tested on observable behaviour. A typical example of a system that corresponds with this approach is the Almost Ideal Demand System (see Deaton and Muellbauer, 1980a).

Still another approach takes a short cut and writes down the demand functions directly in terms of m and $\mathbf{p}$ and a set of coefficients. Byron

(1970), for example, applies the double logarithmic formulation of demand functions, (9), with constant η_i and ϵ^*_{ij}. The homogeneity condition is then (10) which can be tested. This specification is somewhat less convenient if one also wants to test for Slutsky symmetry. Theil's (1965) formulation of the Rotterdam System is in this respect more tractable and at the same time also allows for testing homogeneity. Theil multiplies both sides of (9) in differential form by the expenditure share w_i.

2.3 Sources of violations

2.3.1 Invalid axioms

It may be true that one or more of (A1)–(A6) are invalid, which may lead to a violation of the homogeneity property, (P2). This is sometimes called 'money illusion'. In that case, changes in absolute (rather than relative) prices may not be recognized as such and, therefore, induce changes in real expenditure patterns.

Locating the source of a violation of homogeneity in defective axioms has been a rare response. Marschak (1943), an early tester of homogeneity, is an exception. The problem with this approach is that, apart from 'anything goes', a viable alternative theory of demand does not exist, at least in the micro-economic domain. Deaton (1974, p. 362) is typical. He rejects this response, as it would require an 'unattractive' hypothesis of irrational behaviour.

An alternative that still imples rational behaviour is to add a stochastic element to the utility function (Brown and Walker, 1989). This is an interesting point, as it shows the differences between the object (the observing econometrician) and the subject (the rational individual). Even though the subject behaves consistently with (A1)–(A6), the econometrician cannot observe all relevant factors that determine the individual's preferences. In most empirical studies, this is reflected in adding a disturbance term to demand equations 'as a kind of afterthought' (Barten, 1977). By relating a stochastic disturbance term directly to an individual's utility, it is acknowledged that individuals differ and that these differences are not observable to the econometrician. If the utility function is extended with a disturbance term, the demand equations will have random parameters. The econometrician not only has to stick an additive disturbance term to his specification, but also has to take care of random parameters (as, for example, in Brown and Walker, 1989).

2.3.2 Invalid restrictions

The restrictions (R1)–(R3) above may be invalid. First, consider (R1), linearity of the budget constraint. In reality, prices may be a decreasing function of the quantity bought. It is difficult to think of cases where this invalidates the homogeneity condition directly.

The functional form, (R2), is more likely to generate a clash between theory and empirical results. Since the pioneering work of Schultz, Stone and others, extensive work has been done to create more general functional forms for utility functions or demand equations. In addition, the systems approach which takes advantage of the cross-equation restrictions meant a leap forward. During the 1970s, much effort was spent on inventing more flexible functional forms, inspired by Diewert's (1971) use of duality. It is not clear, however, how the choice of a functional form affects tests of the homogeneity condition (apart from the fact that, for some functional forms, tests of homogeneity cannot be carried out). The increased subtlety of the specifications does not generally result in 'better' test results. More recently, non-parametric approaches to inference have been used to test the theory of consumer behaviour.

The choice of arguments that appear in the demand equation, (R3), may be inappropriate. For example, some variables are omitted. The validity of an empirical test of homogeneity or symmetry is conditional on the correct specification of the model, the maintained hypothesis. Three examples of omissions are:

(a) snob goods (goods that are more desirable if their prices go up, also known as Veblen goods). For such goods, the utility function must include prices. However, such goods are rare and may, therefore, be legitimately omitted. Snob goods have, to the best of my knowledge, never been invoked to explain a rejection of homogeneity;
(b) the planning horizon. If consumers maximize lifetime utility rather than contemporary utility, the optimization problem should be formulated in lifetime terms. However, if m in (3) is sufficiently highly correlated with the life-long concept of expected available means and if the price expectations are also highly correlated with the prices in (3), the included variables take over the role of the omitted ones and homogeneity should hold. The mentioned considerations are an empirical matter, not a purely theoretical one. Hence, homogeneity may or may not hold in the specification. This is a problem that cannot be settled by *a priori* deductive reasoning. In general, mis-specification of (3) might result in biased estimates of the coefficients in the demand equations. The bias could be away from or towards homogeneity. Dynamic mis-specification is not the only possible source of mis-specification. Omission of variables representing changes in preferences can obscure homogeneity even if it exists;
(c) goods not included in the goods bundle considered. In theory, the maximization problem should be formulated in the most general form

possible, with all conceivable arguments that any consumer might be willing to consider. All conceivable qualities of goods should be specified in the choice set. This is impossible for the empirical economist. In applied work, goods are aggregated. The level of aggregation differs but it is impossible to do without aggregation.

2.3.3 Invalid approximations

All empirical demand systems approximate some kind of ideal demand system. It is possible that a rejection of homogeneity is caused by approximation to the ideal rather than truely non-homogeneous demand equations. As Theil (1980, p. 151) remarks, translog specifications (such as employed by Christensen, Jorgenson and Lau, 1975) have been recommended on the ground that quadratic functions in logarithms can be viewed as second-order Taylor approximations to the 'true' indirect utility function. They are accurate when the independent variables take values in sufficiently small regions. Still, even if these regions are sufficiently small, the implied approximation for the demand functions may be very unsatisfactory. It follows from Roy's identity that the demand equations are only first-order approximations (in this respect the indirect translog system is similar to the Rotterdam system).

Clearly, there is a basis for doubt about the validity of the homogeneity condition in a specific model. At the same time, it is not clear how important deviations from the ideal world characteristics which underlie the theory of demand are. Homogeneity, like symmetry, is an ideal world consequence. For several reasons it may be useful to impose it on empirical models of actual behaviour. Whether this imposition is valid, depends not only on the resemblance of reality to the ideal world, but also on the validity of the auxiliary assumptions which are used above. These considerations inspired the investigations of the homogeneity condition in consumer demand.

Another approximation relates to how the variables are measured. For example, m is usually calculated using the budget equation, (R1). The impact this has on the empirical validity of the homogeneity condition is not clear. The measurement error is of relatively little interest, at least empirical researchers have rarely investigated errors in variables as a source of rejecting homogeneity. (Stapleton, mentioned in Brown and Walker, 1989, is an exception but he does not provide sufficient evidence to show that empirical results are driven by measurement error.)

2.3.4 Invalid aggregation

The properties of demand functions for individual agents and elementary goods do not necessarily carry over to demand functions for aggregates of goods by aggregates of individuals. All empirical investigations cited in this chapter deal with such aggregate demand functions. In more recent empirical investigations, applied to cross sectional microeconomic data, homogeneity and symmetry remain occasionally problematic (see the survey by Blundell, 1988).

Conditions for aggregation of *goods* are spelled out in Hicks ([1939] 1946). Most economists are not troubled by aggregation of goods. Empirical deviations from Hicks' composite commodity theorem (if the relative prices of a group of commodities remains unchanged, they can be treated as a single commodity) are thought to be non-essential.

Apart from aggregating goods, economists often test demand theory by aggregating *individuals*. Again, this may be inappropriate. Conditions for aggregating individuals are sketched in Deaton and Muellbauer (1980a, b). An interesting step has been the formulation of an Almost Ideal Demand System (AIDS), that takes care of flexible functional form and aggregation issues. Conditions for aggregation are so strong, that many researchers doubt that findings for aggregate data have any implication at all for the theory of consumer demand that deals with individual behaviour. Blundell (1988) discusses examples of testing the theory using non-aggregated data. Still, even if data for individuals are used, the properties of demand are occasionally rejected by the data.

An additional 'problem' for testing the theory in such a context is that the number of observations tends to be very large. A typical example is Deaton (1990), who estimates demand functions given a sample of more than 14,000 observations. With such large samples any sharp null hypothesis is likely to be rejected given a conventional (5 or 1%) significance level. Not surprisingly, Deaton reports a rejection (of symmetry, homogeneity is not tested). Deaton's resolution to this rejection is to become an expedient Bayesian. He invokes the Schwarz (1978) criterion, which 'saves' the symmetry condition.

2.3.5 Invalid stochastics

The statistical tests reported below all rely on assumptions concerning the validity of the test statistics. Most important are assumptions related to the appropriateness of asymptotic statistics to small samples and assumptions about the distribution of the residuals. These assumptions are frequently doubtful.

Consider the early econometric efforts to estimate demand functions, e.g. Moore (1914). Model specifications were selected after comparing the goodness of fit of alternative models. Karl Pearson's (1900) goodness of fit (χ^2) test was the tool, but the properties of this test statistic (in particular the relevance of degrees of freedom) were not well understood before 1921. Similarly, more recently, systems of demand have been estimated. Testing homogeneity in a system of demand equations is dependent on the unknown variance–covariance matrix. In practice, this has to be replaced by an estimate, usually a least squares residual moment matrix. Econometricians who used these tests were aware that the samples available for investigating consumer demand were not large enough to neglect the resulting bias, but Laitinen (1978) showed that the importance of this problem had been grossly underestimated. Many statistical tests do not have well understood small sample properties – even today. I will return to this issue in section 5 below.

Statistical tests of homogeneity also usually assume specific distributions of the residuals. Models of random utility maximization problem (Brown and Walker, discussed above) suggest that the assumption of homoscedasticity may be invalid. If the residuals are heteroskedastic and no correction is made, tests tend to over-reject homogeneity.

In short, in the econometric evaluation of consumer behaviour, investigators started with two additional 'axioms':

(A7) Axiom of correct specification

(A8) The sample size is infinite

(A7) relates to the validity of the 'maintained hypothesis'. This is rarely the case, taken literally, but econometricians behave as if their model is correctly specified. Still, (A7) may be invalid, causing a rejection of homogeneity. This is related to invalid restrictions (R2) and (R3), but the purely statistical argument about the stochastics of the demand equations has been further exploited by those who favour a non-parametric approach in inference, in which no specific distributional assumptions for the residuals are presumed. Varian (1983) argues that a major drawback of parametric testing is the fact that a rejection can always be attributed to wrong functional forms or mis-specification.

(A8) might also be labelled as the asymptotic idealization. Many tests are based on this assumption of the availability of an infinite amount of data. Again, no one would claim that a particular sample obeys this requirement, but it is a matter of dispute when 'large' is large enough. In econometrics, it is clear that sample sizes are not quite as large as (A8) would require. This was acknowledged early on, for example by Haavelmo (1944) who relies on a weakened version of (A8):

(A8′) The sample is a random drawing from a hypothetical infinite population.

As shown in chapter 6, this version of Asymptopia has gained tremendous popularity among econometricians, but an uncritical belief in (A8′) may result in an unwarranted rejection of the properties of demand.

3 Empirical work

The history of testing the homogeneity condition reads as a short history of econometric inference. All kinds of problems of inference are discussed, different approaches in statistics are applied to estimating and testing demand equations. In this section, I use a number of empirical studies to illustrate some of the problems of econometric inference.

3.1 Early empirical work

3.1.1 Goodness of fit

The earliest known statement about the law of demand goes back to Davenant (1699) who observed a negative relation between the quantity of corn and its price. The relation differs from (3) in several respects. First of all, it takes the quantity as given and the price to be determined. This makes sense for agricultural goods. For the purpose at hand, the fact that only one good and one price are considered is more important because homogeneity involves a set of prices and means. The simplified relation may be justified by an appeal to a *ceteris paribus* clause: the other prices and means are taken to be constant. Whether this can be maintained empirically was beyond the scope of Davenant. Most likely, the assumption was invalid, which does not diminish the importance of his effort as a crude first approximation.

Davenant's assumption was used unchanged until the nineteenth century, when the quantity demanded was also made dependent on prices other than its own. Walras ([1874–7] 1954) needed all quantities to depend on all prices and *vice versa* for his general equilibrium framework. According to Walras, the choice of the *numéraire*, the unit in which the prices are expressed, is arbitrary. This, of course, is a way of stating the homogeneity condition. Walras did not pursue an empirical analysis. Henry Moore was one of the first economists to make a statistical investigation of demand functions. Homogeneity was not among his interests, however.

What is of interest in Moore's work is the source of his statistical theory. Moore attended courses in mathematical statistics (1909) and correlation (1913) with Karl Pearson (Stigler, 1962). Pearson, one of the champions of positivism and measurement, was a clear source of inspiration in Moore's efforts to measure elasticities of demand. Two of Pearson's innovations, the correlation coefficient and the goodness of fit test, were instrumental in Moore's investigations. Stigler notes that Moore was not aware of losses of degrees of freedom in his applications of the χ^2 test. This is no surprise as the issue was not solved before 1921 when Fisher provided the argument. Pearson himself was most reluctant to accredit this point to Fisher (see Fisher Box, 1978, pp. 84–5). Moore's use of goodness of fit tests was as modern as could be expected.

Moore's legacy not only consists of his early empirical work on demand, but is also his 'professed disciple' (Stigler, 1962), Henry Schultz. Schultz (1928), a pioneering authority in this field, estimates demand functions for sugar. After a short discussion of Cournot and Marshall (suggesting a demand equation with the demanded good's own price as the only explanatory variable; a *ceteris paribus* clause states that all other prices are constant), Schultz turns to the so-called 'mathematical school' of Walras and Pareto. Here, the quantity demanded is a complicated function of many price variables. Schultz (p. 26) asks:

> How can we deal with such complicated functions in any practical problem? The answer is that, although *in theory* it is necessary to deal with the demand function in all its complexity in order to show that the price problem is soluble, *in practice* only a small advantage is gained by considering more than the first few highly associated variables.

Schultz follows the pragmatic approach, by estimating a simple sugar demand equation with real income and prices deflated by the consumer price index as explanatory variables. His motivation is neither to exploit homogeneity explicitly, nor to test it, but to see if the fit is better than when using absolute prices.

Schultz (1937) makes use of theoretical insights of Slutsky and Hotelling and even provides an informal test of the symmetry condition (pp. 633, 645). A formal test for homogeneity is not pursued, although some casual information on its validity is provided. After discussing the question of whether estimation of demand functions should use absolute or relative prices, Schultz (p. 150) chooses the latter, because 'competent mathematical statisticians have long felt that every effort should be made to reduce the number of variables in a statistical equation to a minimum'.

The choice of variables is inspired by this pragmatic statistical motivation, not by theoretical argument. The informal test is to compare a demand equation using real price and consumption *per capita* on the one hand, and nominal price and nominal consumption *per capita* on the other hand. Scaling by a price index and a population index has the same goal: to improve the fit. The results (p. 71 and p. 80) show no difference in fit.

3.1.2 *Money illusion?*

Schultz is not interested in testing theory or the homogeneity condition. Rather, the purpose is to estimate elasticities. Marschak (1943) goes beyond this: he wants to test the null hypothesis of 'absence of money illusion'.[3] The following motivation is given by Marschak (p. 40):

> For the economist, our 'null hypothesis' has an additional interest on the grounds (1) that it is a necessary (though not sufficient) condition of rational behavior, defined as using one's income to one's best satisfaction, (2) that it supplies a justification for using 'deflators' in economic statistics and for discussing the demand relationships in terms of so called 'real' incomes and prices, and (3) that it is incompatible with important theories of unemployment.

Two incompatible theories mentioned are Irving Fisher's theory that people are more sensitive to changes in income than to changes in prices, and Keynes' theory of unemployment. The theoretical background is the Slutsky equation from which the restrictions on demand functions are derived. Marschak's empirical model of demand is a combination of an Engel curve and price substitution effects. Just one demand equation is estimated: demand for meat. The explanatory variables are income and the prices of meat, other food and non-food goods. The test carried out is not a formal significance test, but standard errors of the estimates of elasticities are given. Using (7) for the test, the estimated elasticities reject Marschak's null hypothesis: the numerical magnitude of the income elasticity is larger than the sum of the price elasticities. Marschak is almost alone in the history of testing homogeneity for his willingness to interpret his rejection as a falsification, although a 'sweeping verdict' on the validity of the null hypothesis should not yet be made (p. 48).

3.1.3 *Stone and the measurement of consumer behaviour*

The next important study is Stone (1954a). His model is an example of the set of double logarithmic demand equations, discussed above. He (p. 259) employs (10) for the actual test. Estimates are given for thirty-seven categories of food, based on data for the period 1920–38. The equations are parsimonious (not all substitution terms are included – this would be impossible, given the limited number of observations). Although they result from a search for 'the best significant equation' (p. 328), no one would regard Stone's efforts as thoughtless data-mining. He (p. 254) gives the following motivation for his investigation of demand theory:

> In the first place it gives rise to a number of hypotheses about the characteristics of demand equations which, in principle at any rate, are capable of being tested by

an appeal to observations. These are the 'meaningful theorems' of Samuelson, which, however, will here be termed verifiable theorems. One such theorem is, for example, that consumers do not suffer from 'money illusion' or, put more specifically, that demand equations are homogeneous of degree zero in incomes and prices so that a simultaneous multiplication of incomes and prices by the same positive constant would leave the amounts demanded unchanged.

In the second place these implications of consistent behaviour usually involve some restriction on the parameters of the demand equations. Thus the homogeneity theorem just referred to entails that the sum of the elasticities of demand with respect to income and all prices is zero. (p. 254)

Stone refers to Samuelson ([1947] 1983, p. 4), who defines a meaningful theorem as 'a hypothesis about empirical data which could conceivably be refuted, if only under ideal conditions'. From the actual test, Stone (1954a, p. 328) infers that,

with two exceptions, home produced mutton and lamb, and bacon and ham, the sums of the unrestricted estimates of the price (substitution) elasticities do not differ significantly from zero. Accordingly, the assumption that the condition of proportionality is satisfied is supported by the observations.

The significance level employed is the familiar 5%. Stone wants to verify, not falsify. Had he been a falsificationist, he would have concluded that the two rejections cast demand theory in doubt, as the theory should apply to all demand equations together. Instead, he appears to be pleased with the results and continues by imposing homogeneity, interpreting the estimated equations as proper demand functions, and the parameters as proper elasticities.

3.2 Empirical work based on demand systems

The first test of homogeneity in a systems approach is carried out in Barten (1967). The test is a joint one for homogeneity and symmetry, hence not a test of homogeneity alone. The homogeneity part of the test is based on (10). Although it was calculated, the result of a separate test for homogeneity was not published. This result suggested that the homogeneity condition was not well supported by the data, but it was unclear whether the result was robust. The joint test of homogeneity and symmetry, which was published, did not lead to strong suspicion of these conditions taken together. Testing demand theory is not the main goal of the paper: the focus is on estimation.

In this context, and given the very strong *a priori* appeal of the homogeneity condition, Barten summarizes the evidence on homogeneity as follows:

we may conclude that the empirical data used here are not inconsistent with the homogeneity property. That is, we have not been able to find with any great precision the effect of a monetary veil. (p. 81)

However, the first published formal test of homogeneity in a systems approach, Barten (1969), is a rejection of homogeneity. The model uses the Rotterdam specification for sixteen consumption categories and data (for the Netherlands) ranging from 1922–39 and 1949–61. It is estimated with maximum likelihood. Applying maximum likelihood estimation to a complicated problem, showing its fruitfulness, is the real purpose of this paper. An easy by-product is the application of a likelihood ratio statistic to test homogeneity.

Hence, testing homogeneity itself is not the major goal of the paper – but the result was as unexpected as disappointing (the disappointment may have stimulated efforts to counter the rejection, making it a source of inspiration for the subsequent specification search). The rejection confirmed the unpublished earlier one: homogeneity seemed to be on shaky grounds, although a viable alternative was not considered. The basic motivation for testing was that it could be done: it showed the scope of the new techniques.

The strong *a priori* status of the homogeneity condition made Theil suggest not publishing the rejection, as this might have been due to computational or programming errors. Publication would unnecessarily disturb the micro-economics community. When it became clear that the results were numerically correct and the rejection was published, responses varied. Some authors argued that demand systems which invalidate homogeneity do not count as interesting ones (e.g. Phlips). Others working with different specifications and with different estimation and testing techniques tried to replicate the experiment.

Byron (1970) replicates using the double-logarithmic approach to postulate a demand system. Using the same data as Barten (1969), Byron duplicated the rejection of homogeneity (and symmetry). After providing his statistical analysis, Byron (1970, p. 829) finds it 'worthwhile to speculate why the null hypothesis was rejected'. He continues:

It is possible that the prior information is simply incorrect – that consumers do not attempt to maximize utility and do experience a money illusion due to changing prices and income. It is quite likely, however, that this method of testing the hypothesis is inappropriate, that aggregation introduces errors of a non-negligible order of magnitude, that the applications of these restrictions at the mean is excessively severe for data with such a long time span, that the imposition of the restrictions in elasticity form implies an unacceptable form of approximation to the utility function, that preferences change, or that the naive utility maximization hypothesis needs modification for the simultaneous consideration of perishables and durables. Such arguments are purely speculative: the only definite information available is that the imposition of the restrictions in this form on this data leads to the rejection of the null hypothesis.

In other words, Byron lists a number of auxiliary hypotheses that may cause the rejection. The tested hypothesis itself may be right if one of the auxiliary hypotheses in the test system is wrong. This is the Duhem–Quine thesis, discussed in chapter 1, section 5.1. A theory is an interconnected web, no test can force one to abandon a specific part of the web (Quine, [1953] 1961, p. 43). I consider this issue on page 203 below. The problem for Barten, Byron and other researchers is that they fail to have a preconception of where to go: they do not know what lesson to learn from the significance tests.

The rejection of homogeneity for Dutch data might be an artifact of the Netherlands (although not all tests on Dutch data yielded a clear rejection). Hence, other data sets were used for estimating and testing a systems approach to consumer demand. The homogeneity condition was rejected for Spanish data using both the Rotterdam specification and the double-logarithmic one by Lluch in 1971. The UK is next in the search for homogeneity. Deaton (1974) compares a number of specifications of demand systems (five versions of the Rotterdam system, as well as the LES, Houthakker's direct addilog system and a model without substitution effects). The data range from 1900 to 1970 and distinguish nine consumption categories for the UK. Using the likelihood ratio test, Deaton (p. 362) concludes that the null hypothesis of homogeneity is firmly rejected for the system as a whole. A closer look at the individual demand equations shows that (P2) cause the problems. This result is consistent with the observation in Brown and Deaton (1972, p. 1155):

Of the postulates of the standard model only one, the absence of money illusion, has given consistent trouble; there is however some evidence to suggest that this result can be traced to individual anomalies.

These results illustrate the failure to confirm homogeneity starting from the direct postulation of demand functions. An example of the alternative, to start from indirect utility, is Christensen, Jorgenson and Lau (1975). On the basis of rejecting symmetry, conditional on homogeneity, they (p. 381) conclude that 'the theory of demand is inconsistent with the evidence'. Their evidence adds new information to previous results. According to the authors, it might be that demand theory is valid, but that utility is not linear logarithmic. But the results 'rule out this alternative interpretation and make possible an unambiguous rejection of the theory of demand'.

The continuous stream of publications based on the theory of demand makes clear, however, that few economists did or do share this conclusion of Christensen, Jorgenson and Lau, even though the sequence of rejections of the homogeneity condition does not stop here. A novel effort to test homogeneity is made by Deaton and Muellbauer (1980a), who apply the Almost Ideal Demand System using British data from 1954 to 1974 and eight consumption categories. F-tests on the restriction of homogeneity reject homogeneity in four of the eight cases. Because the Durbin–Watson statistic indicates problems in exactly the same cases, Deaton and Muellbauer conjecture that the rejection of homogeneity is a symptom of dynamic mis-specification. Here, a combination of tests points to where the investigator might search. But this direction is not pursued in Deaton and Muellbauer (1980a).

4 Philosophy

4.1 Rejection without falsification

Samuelson's *Foundations* aims at deriving 'meaningful theorems', refutable hypotheses. The economic theorist should derive such theorems, and 'under ideal circumstances an experiment could be devised whereby one could hope to refute the hypothesis' (Samuelson, [1947] 1983, p. 4). This sounds quite Popperian (*avant la lettre*, as the English translation of *The Logic of Scientific Discovery* appeared much later). But at the same time, Samuelson is sceptical about testing the theory of consumer behaviour, as revealed by the opening quotation of this chapter. It is hard to be Popperian.

This is also the conclusion of Mark Blaug (1980, p. 256), who argues that much of empirical economics is

> like playing tennis with the net down: instead of attempting to refute testable predictions, modern economists all too frequently are satisfied to demonstrate that the real world conforms to their predictions, thus replacing falsification, which is difficult, with verification, which is easy.

Following Popper ([1935] 1968) and Lakatos (1978), Blaug values falsifications much more highly than confirmation. Economists should use every endeavour to try to falsify their theories by severe tests. Blaug (1980, p. 257) complains that economists do not obey this methodological principle, and stoop to 'cookbook econometrics', i.e. they 'express a hypothesis in terms of an equation, estimate a variety of forms for that equation, select the best fit, discard the rest, and then adjust the theoretical argument to rationalize the hypothesis that is being tested'.

It is doubtful that Blaug's general characterization of empirical economics holds water. Of course, there are poor econometric studies but, in order to criticize, one should also consider the best contributions available. The search for homogeneity certainly belongs to the best econometric research one can find. The case of homogeneity is illuminating, as the homogeneity condition makes part of a well-defined economic theory which provides the empirical researcher a tight straitjacket. Data-mining or 'cookbook econometrics' is ruled out.

The empirical investigations discussed above show how the game of investigating consumer demand was played: with the net up. At various stages in time, the most advanced econometric techniques available were used. But there is a huge difference between playing econometrics with the net up, and falsifying theories. Few researchers concluded that the theory of consumer demand was falsified. Verification of homogeneity turned out to be difficult, not easy.

One rejection might have induced a ('naive') Popperian to proclaim the falsification of the neoclassical micro-economic theory of consumer behaviour. But such a 'crucial test' is virtually non-existent in economics. Blaug ([1962] 1985, p. 703) warns, 'conclusive once-and-for-all testing or strict refutability of theorems is out of the question in economics because all its predictions are probabilistic ones'. But a sequence of rejections should persuade more sophisticated adherents of fallibilism. Lakatosians might conclude that the neoclassical research programme was degenerating. In reality, none of these responses obtained much support. Instead, a puzzle emerged, waiting for its solution. Homogeneity was rejected, not falsified. This stubbornness is hard to justify in falsificationist terms.

Do Marschak and Christensen, Jorgenson and Lau qualify as Popperians? Christensen *et al.* certainly not, as a repeating theme in Popper's writings is that theories can only be falsified if a better theory, with a larger empirical content, is available (see also Blaug, [1962] 1985, p. 708). Such an alternative is not proposed. This point is sometimes overlooked by statisticians, although most are aware of this limitation of significance testing. According to Jeffreys ([1939] 1961, p. 391), the required test is not whether the null must be rejected, but whether the alternative performs better in predicting (i.e. is 'likely to give an improvement in representing future data'). Surprisingly perhaps, such a test has not been carried out in the investigations of consumer demand presented above (more recently, this gap has been filled; see the discussion of Chambers, 1990, below). How, then, should we interpret the bold claim made by Christensen, Jorgenson and Lau (1975) that the theory of demand is inconsistent with the evidence? They may have been under the spell of falsificationism (perhaps unconsciously, they do not refer to any philosophy of science). The Popperian tale of falsification was popular among economists during the 'seventies (and alas still is). But if Popper's ideal is praised, then the praise is not very deep. An example is Deaton (1974, p. 345) :

> A deductive system is *only* of practical significance to the extent that it *can* be applied to concrete phenomena and it is too easy to protect demand theory from empirical examination by rendering its variables unobservable and hence its postulates unfalsifiable.

After embracing the falsificationist approach, and after finding that homogeneity is on shaky grounds, Deaton (p. 362) remains reluctant to conclude to a falsification of demand theory.

4.2 Popper's rationality principle

Now consider Marschak. He tests homogeneity and has an alternative: existence of money illusion (or more generally, a Keynesian model of the economy although this is not specified). This seems to be close to the Popperian ideal; but there is a problem. Popper is aware that, in the social sciences, the testability of theories is reduced by the lack of natural constants (e.g. Popper, [1957] 1961, p. 143). This might diminish the importance of the methodology of falsificationism for the social sciences (this is the position of Caldwell, 1991; see also Hands, 1991).

On the other hand, an advantage of the social sciences relative to the natural sciences is the element of rationality (Popper, [1957] 1961, pp. 140–1). Popper proposes supplementing methodological falsificationism with the 'zero method' of constructing models on the assumption of complete rationality. Next, one should estimate the divergence between actual behaviour and behaviour according to the model. Marschak, however, not only measures the divergence but constructs a test of rationality. Popper is ambiguous on this point: it is unclear whether methodological falsificationism extends to trying to falsify the rationality principle. The fact that Popper refers to Marschak in favourable terms suggests that falsificationism is the primal principle, with the rationality principle second.

Whatever the verdict will be on the Popperian spirits of Marschak, he has not changed the general attitude towards demand theory in general or money illusion specifically. Stone and most other investigators of consumer demand continue to work with neoclassical demand theory. Even if Marschak were Popperian, his successors in economics are not.

It has been noted by philosophers of science that actual science rarely follows Popper's rules (Hacking, 1983; Hausman, 1989). In particular, Lakatos has tried to create a methodology that combines some of Popper's insights with the actual proceedings of science. Can we interpret the history of testing homogeneity in Lakatos' terms?

4.3 Does Lakatos help?

Using Lakatos' (1978) terminology, the homogeneity condition belongs to the 'hard core', the undeniable part of the neoclassical scientific research programme, together with statements like 'agents have preferences' and 'agents optimize'. Some economists argue that demand equations are meaningless if they violate the homogeneity condition. The hard core status of homogeneity is clear in the writings of Phlips (1974, p. 34): functions that do not have this property do not qualify as demand equations.

The essential characteristic of a hard core proposition is that a supporter of the research programme should not doubt its validity. The hard core proposition is irrefutable by the methodological decision of its proponents. It is an indispensable part of a research programme. At first sight, such an interpretation of the homogeneity postulate seems to be warranted: see Samuelson's epigraph to this chapter and the quotations of Marschak (1943, p. 40) and Stone (1954a, p. 254). Deaton (1974, p. 362) concurs:

> Homogeneity is a very weak condition. It is essentially a function of the budget constraint rather than the utility theory and it is difficult to imagine any demand theory which would not involve this assumption. Indeed to the extent that the idea of rationality has any place in demand analysis, it would seem to be contradicted by non-homogeneity.

Hence, Deaton claims that any demand theory based upon rational behaviour of consumers should satisfy the homogeneity condition: accepting the rejection 'implies the acceptance of non-homogeneous behavior and would seem to require some hypothesis of "irrational" behavior; this is not an attractive alternative'.

If we continue with the attempt to use Lakatos for an understanding of the analysis of homogeneity, we must introduce the 'negative heuristic' (see chapter 1, section 5.2), the propositions which immunize the hard core. In our case, the negative heuristic is not to test the hard core proposition of homogeneity. But this picture clearly does not fit reality. Empirical investigators were eager to test homogeneity. They did exactly what this supposed negative heuristic forbade.

Hence, either homogeneity changed status, from hard core proposition to something else, or the Lakatosian scheme is not suitable for an understanding of the history of consumer demand analysis. The first option, a change in status, would imply that homogeneity became part of the so-called 'protective belt of auxiliary hypotheses which has to bear the brunt of tests and get adjusted and re-adjusted, or even completely replaced, to defend the thus-hardened core' (Lakatos, 1978, p.48). The protective belt consists of the findings that may be used to support the hard core propositions, or which may be investigated as puzzles, anomalies.[4] But the testers of homogeneity listed many hypotheses that might explain the rejection of homogeneity. Byron's list, cited above, is one example. The auxiliary hypotheses are related to dynamics, parameter drift, small samples, etc., but homogeneity itself is not considered as yet another auxiliary hypothesis to the hard core of demand theory.

Lakatos' classification is of little help. And if this classification cannot be used, it is not fruitful to go on with a rational reconstruction of the history of research in consumer demand using the methodology of scientific research programmes. There is no base for concluding that neoclassical demand theory was either degenerative or progressive. The Lakatosian philosophy of science leaves us empty handed. This observation corresponds to the view of Howson and Urbach (1989, p. 96), who argue that Lakatos was unable to clarify what qualifies elements of a theory to become part of the hard core of a research programme. Lakatos invokes a 'methodological fiat', but what is this other than pure whim? Unfortunately, Howson and Urbach write,

> this suggests that it is perfectly canonical scientific practice to set up any theory whatever as the hard core of a research programme, or as the central pattern of a paradigm, and to blame all empirical difficulties on auxiliary theories. This is far from being the case.

It is hard to provide a coherent Popperian or Lakatosian interpretation of this episode in the history of testing: we observe a sequence of rejections, without falsification. Does this mean that 'anything goes' (Feyerabend, 1975)?

4.4 Auxiliary hypotheses and the Duhem–Quine thesis

The answer to the question at the end of the previous subsection is no. The tests suggest that something is wrong in the testing system. Quine (1987, p. 142) recommends using the maxim of minimum mutilation: 'disturb overall science as little as possible, other things being equal'. Dropping homogeneity is far more disturbing than dropping, for example, time-separability of consumption, homoskedastic errors, or the 'representative individual'. Dropping homogeneity, i.e. studying consumer behaviour without demand theory (or a viable alternative), would make the measurement of elasticities very complex.

Research was directed to saving the straitjacket, the attack directed at the auxiliary hypotheses. Recall the sources of violation of homogeneity, other than failure of demand theory (or the axioms underlying the theory) itself. They are invalid restrictions, invalid approximations, invalid aggregation and invalid stochastics.

Linearity of the budget restriction, (R1), has not been questioned by the demand researchers. It has fairly strong empirical support, deviations from linearity are small and, moreover, it is hard to imagine how these could cause non-homogeneity. The choice of functional form, (R2), on the other hand, is crucial. In some cases (the LES, for example) homogeneity is simply imposed. In other cases, it can be imposed by restricting the parameters of the demand equation. The functional form of the demand equation is open to much doubt. It is widely thought that parametric demand equations are only rough approximations to some ideal demand function. Hence, the 'truthlikeness' or *a priori* probability of the specification is relatively low, and much research has been directed to obtaining better approximations to this ideal. The same applies to the choice of arguments, (R3). The variable 'time' has received special attention (Deaton and Muellbauer, 1980a). Other ones, such as snob goods, are thought to be innocuous and have not attracted research efforts to save homogeneity.

Aggregation is of special interest. According to some, there is no reason to believe that the micro relations (such as the law of demand) should hold on a macro level. Others claim that economic theory can only be applied to aggregates. This issue inspired research into the conditions for aggregation, in particular of individuals (rather than goods). The effects of income distribution are crucial. Still, even in aggregate demand systems where care is taken of income effects homogeneity is violated. The same happens occasionally with disaggregated demand studies.

Finally, the particular validity of the statistical tests that were employed has been questioned but it was not until the late seventies that this approach convinced applied econometricians of the fact that homogeneity, after all, could stand up against attempts to reject it if the right small sample tests were used. The old results could be given a new interpretation; a kind of *Gestalt* switch had taken place. It is interesting that the validity of statistical tools is a crucial auxiliary assumption that turned out to be a false one. I will return to this in the next section.

5 The theory versus the practice of econometrics

5.1 Statistical testing using frequency concepts

If we reconsider the dominant approaches in frequentist probability theory, then Fisher's seems to be the most appropriate in the context of the analysis of consumption behaviour. In most of the studies presented in this chapter, one cannot speak of a von Mises collective. Von Mises explicitly disavowed small sample analysis. It is only more recently that econometricians have started to analyse consumer behaviour by means of micro (cross section) data. Deaton (1990) is an example. However, these recent investigations are not inspired by a quest for collectives, but by the fact that the theory of consumer behaviour is about individuals, not aggregates.

Fisher's (1922, p. 311) 'hypothetical infinite population, of which the actual data are regarded as constituting a random sample' is the metaphor by which the applied econometricians could salve their frequentist consciences, even though the method of maximum likelihood does not enter the scene before the late sixties. The significance tests are interpreted as pieces of evidence, although in most cases it is not clear what evidence. But a number of differences from Fisher's approach stand out. Initially, the aim of inference is not the reduction of data: the real aim is measurement of elasticities, which do not have to be sufficient statistics. Second, fiducial inference is not a theme in demand analysis. Third, there is no experimental design or randomization by which parametric assumptions could have been defended and specification uncertainty reduced.

Is the Neyman–Pearson approach closer to econometric practice? Not quite. First, the decision context is not clear. Why would the econometrician bother with making a wrong decision? This critique is anticipated by Fisher, who argues that in the case of scientific inference, translating problems of inference to decision problems is highly artificial.

Second, the Neyman–Pearson lemma is of little relevance in the case of testing composite hypotheses, based on what most likely is a mis-specified model, using a limited amount of data without repeated sampling. In these circumstances, the properties of the test statistics are not well understood. Indeed, the Neyman–Pearson framework has hardly been relevant for testing the properties of demand, although references to 'size' and 'power' pervade the literature.

Third, although the Neyman–Pearson approach stands out for its emphasis on the importance of alternative hypotheses, there are no meaningful alternatives of the kind in the case of testing homogeneity. Thus, as Jeffreys ([1939] 1961, p. 390) wonders: 'Is it of the slightest use to reject a hypothesis until we have some idea of what to put in its place?'

Fourthly, homogeneity is a singularity whereas the alternative is continuous. If enough data are gathered, it is most unlikely that the singularity precisely holds (Berkson, 1938). This issue is recognized by Neyman and Pearson. They introduce their theory of testing statistical hypotheses with a short discussion of a problem formulated by Bertrand: testing the hypothesis that the stars are randomly distributed in the universe. With this in mind, they state (1933a, pp. 141–2) that

> if x is a continuous variable – as for example is the angular distance between two stars – then any value of x is a singularity of relative probability equal to zero. We are inclined to think that as far as a particular hypothesis is concerned, no test based upon the theory of probability can by itself provide any valuable evidence of the truth or falsehood of that hypothesis.

Their proposal to circumvent this problem is to exchange inductive inference for rules of inductive behaviour. The problems with that approach have been mentioned in chapter 3.

5.2 The lesson from Monte Carlo

An important approach to attack the puzzle of rejections of homogeneity comes from Laitinen (1978), who shows that the test statistic has properties different from what was believed. One of the problems of testing homogeneity is the divergence between asymptotic theory and small sample practice. (A8) may be invalid. Early on, the use of small samples was recognized as a possible explanation for the rejection of homogeneity (Barten, 1969, p. 68; Deaton, 1974, p. 361). However, Deaton remarks that the small sample correction which has to be made in the case of testing restrictions on parameters within separate equations, such as homogeneity, is known. The correction was generally thought to be too small to count for an unwarranted rejection of homogeneity. Laitinen (1978, p. 187) however claims that the rejections of homogeneity are due to the fact that 'the standard test is seriously biased toward rejecting this hypothesis'. The problem is that testing homogeneity for all equations is dependent on the unknown variance–covariance matrix, which in practice is usually replaced by a least squares residual moment matrix.

Laitinen describes a Monte Carlo experiment of an artificial statistical computer simulation which shows that estimating demand systems with only a few degrees of freedom should make a small sample correction for the test statistic. Particularly in very small samples, the correction factor was much larger than previously thought. This issue was resolved once and for all by the small sample statistic proposed by Laitinen. Applying this has led to less frequent rejections of homogeneity (although rejections still occur).

5.3 The epistemological perspective

After rejecting Popperian or Lakatosian rational reconstructions of the search for homogeneity, and showing how frequentist probability interpretations are hard to reconcile with the efforts to study consumer behaviour, a better interpretation should be given. A quasi-Bayesian approach may do (a formal quantification of the history of evidence is a sheer impossible task, if only because it is hard to interpret the frequentist statistical tests in an unambiguous Bayesian way). One advantage of the Bayesian approach is that there is no longer any need for an artificial distinction between 'hard core' and 'protective belt'. Second, Bayesians value rejections as highly as confirmations: both count as evidence and can, in principle at least, be used for a formal interpretation of learning from evidence.

The distinction between a basic and an auxiliary hypothesis can be refined in a Bayesian framework. Let S denote the hypothesis that a particular specification is valid (for example the Rotterdam system), let H be the 'meaningful theorems' (homogeneity, in our case) and D the data. Although in most cases listed above, the evidence rejects the joint hypothesis or test system, $H \& S$ the effect on the probabilities $P(H|D)$ and $P(S|D)$ may diverge, depending on the prior probability for H and S respectively. Homogeneity has strong *a priori* appeal, for example, due to experiences with currency reform (such as in France) or introspection. The particular specification is less natural. As a result, rejection of the test system will bear more strongly on the specification than on the hypothesis of interest, homogeneity (see also Howson and Urbach, 1989, p. 99). The posterior probability for homogeneity will decline only marginally, as long as the specification is much more doubtful than homogeneity itself. This observation explains why the rejections by a sequence of statistical tests were not directly interpreted as a falsification of homogeneity. If rejections would have dominated in the long run, whatever specification or data had been used, eventually the rational belief in homogeneity itself would have declined.

The homogeneity condition is, for some investigators, so crucial that it should be imposed without further question. A researcher in the frequency tradition would invoke the 'axiom of correct specification', a Bayesian may formulate a very strong prior on it, equal to 1 in the extreme case. If this were done, one would never find out that the model specification is wanting. The specification search for homogeneity is of interest because it violates both the frequentist and the Bayesian maxim.

Following the theory of simplicity, a trade-off should be made between goodness of fit and parsimony in demand systems. The trade-off is goal-dependent. If measurement of elasticities is the real goal of inference (for economic policy evaluation, for example), it may be right to impose homogeneity (otherwise, it is not clear what the meaning of the parameters is). If the research is, first of all, of academic interest, homogeneity should not be imposed except if the resulting model outperforms a model that violates homogeneity given specific criteria (such as predictive qualities). An approach, broadly consistent with the Minimum Description Length principle presented in chapter 5, section 2.4, is chosen in Chambers (1990). Using time series data (108 quarterly observations), he compares different demand systems by means of their predictive performance (using a root mean squared prediciton error criterion). The models are the LES, a LES supplemented with habit formation (introducing dynamics), the AIDS, an error correction system, and a Vector Autoregressive model. Chambers finds that a relatively simple specification, i.e. the LES supplemented by habit formation, outperforms more complicated specifications, such as the AIDS. In the favourite model, homogeneity and symmetry are not rejected by statistical tests.

5.4 In search of homogeneity

Falsification was apparently not the aim of testing homogeneity, but to many researchers the real aim was very vague. Most of the researchers of consumer demand tend to interpret the statistical test of homogeneity as a specification test rather than intentional efforts to test and refute an economic proposition (although this is never explicitly acknowledged and is quite probably only an unconscious motive). Gilbert (1991a, p. 152) makes a similar point. If homogeneity is immune or nearly so for statistical tests, why then take the trouble to write sophisticated software and invent new estimation techniques or specifications in order to test this condition?

Some insights of Leamer (1978) (rarely exploited in methodological investigations) may help to answer the question. Leamer interprets empirical econometrics as different kinds of 'specification searches'. The empirical investigation of the homogeneity condition can be understood as a sequence of different kinds of specification searches. Most tests were neither Popperian efforts to falsify a theory, nor Lakatosian attempts to trace a 'degeneration' of consumer demand theory, but specification tests that served a number of different goals. Leamer

(p. 9) argues that it is important to identify this goal, the type of specification search, as the effectiveness of a search must be evaluated in terms of its intentions. A bit too optimistic, Leamer claims that it is 'always possible for a researcher to know what kind of search he is employing, and it is absolutely essential for him to communicate that information to the readers of his report'. With the search for homogeneity in mind, a more realistic view is that most research workers have only a vague idea of what is being tested. Interpreting the episode with hindsight, we observe that the goal of the empirical inference on homogeneity gradually shifts. Schultz and his contemporaries are searching for models as simple as possible. The significance of their 'test' of homogeneity is the effort to reduce the dimension of the model, which Leamer classifies as a 'simplification search'. A simplification search results from the fact that, in theory, the list of explanatory variables in a demand equation is extremely long, longer than any empirical model could adopt. Particularly during the pre-World War II era, gaining degrees of freedom was important. This simplification search has little to do with a metaphysical view that 'nature is simple' and much more with pragmatic considerations of computability, (very) small sample sizes and efficient estimation.

The theme recurs with Phlips (1974, p. 56) who, writing on imposing homogeneity, holds that 'the smaller the number of parameters to be estimated, the greater our chances are of being able to derive statistically significant estimates'. Marschak's effort to test homogeneity has partly the same motivation, but also arises from his interest in the validity of two rival theories: neoclassical (without 'money illusion') and Keynesian (with 'money illusion'). Although Marschak (like the other demand researchers discussed in this paper) does not use Bayesian language, we may argue that he assigns a positive prior probability to both models and then tests them. This is, in Leamer's terminology, a 'hypothesis testing search', the kind of search that some philosophers of science emphasize. An obvious reason why Marschak's rejection of homogeneity is not taken for granted by his contemporary colleagues (or their successors) is that the large implicit prior against the homogeneity condition of consumer demand was not shared. Furthermore, a general problem with the tests of homogeneity is a lack of consideration of the economic significance of the deviation between hypothesis and data. The interest in such a kind of importance testing has declined while formal statistical significance testing has become dominant. This development is deplored by Varian (1990), who argues that economic significance is more interesting than statistical significance. Varian suggests developing money-metric measures of significance.

Stone, in his book (1954a, p. 254), is motivated by a combination of increasing degrees of freedom (simplification search) and testing the 'meaningful theorems' of demand theory. But with regard to the latter, he does not provide an alternative model to which a positive prior probability is assigned. Hence, it is difficult to interpret Stone's work as a 'hypothesis testing search' similar to Marschak's. The same applies to Schultz's successors. Their attitude is better described as an 'interpretive search'. Leamer defines this as the small sample version of a hypothesis testing search, but it may be better to define it as an instrumentalistic version of hypothesis testing. The interpretive search then lets one act as if one or another theory is 'true'. Take, as an illustration, Deaton's (1978) work. Two specifications (the LES and a PIGLOG specification) of consumer demand are tested against each other, using Cox's test. The disturbing result is that, if either the LES or the PIGLOG specification is chosen as the null, the other specification is rejected. The statistical tests leave us empty handed. Homogeneity is not investigated, but Deaton makes an interesting general remark about this implication of testing, which reveals his instrumentalistic attitude:

> This is a perfectly satisfactory result . . . In this particular case there are clearly better models available than either of those considered here. Nevertheless we might still, after further testing, be faced with a situation where *all* the models we can think of are rejected. In this author's view, there is nothing to suggest that such an outcome is inadmissible; it is perfectly possible that, in a particular case, economists are not possessed of the true model. If, in a practical context, *some* model is required, then a choice can be made by minimizing some appropriate measure of loss, but such a choice in no way commits us to a belief that the model chosen is, in fact, true. (1978, p. 535)

Another type of Leamer's specification searches is the 'proxy search', which is destined to evaluate the quality of the data. Although this issue is raised in a number of the studies discussed, it has not played a major role in practice. Instead, the issue of the quality of the test statistics turned out to be more interesting.

This brings us to Laitinen. His work invalidates many of the rejections of the homogeneity condition that were found during the search for homogeneity. He points to judgemental errors that were made in the interpretation of significance levels. This result is different from the variety of specification searches and needs another interpretation. A Bayesian would argue that the choice of significance levels is a weak cornerstone of classical (frequentist) econometrics anyway, and that these tests are useful only insofar as they refer to a well specified loss structure. For the frequentist econometrician who believes strongly in homogeneity, and worships the significance of testing statistical hypotheses, of Laitinen's result solves a puzzle that is otherwise hard to explain. The salvage is not complete: occasionally, 'strange' test outcomes do occur.

6 Summary

The popular attitude towards testing in econometrics is one of disappointment and scepticism. Blaug (1980) exhibits disappointment. McCloskey (1985, p. 182), who argues that no proposition about economic behaviour has yet been overturned by econometrics, represents the sceptical attitude to significance testing. This scepticism is shared by Spanos (1986). Blaug, who may be characterized as a dissatisfied Popperian, proposes to test better and harder. Spanos (like Hendry) suggests that significance testing (in particular, mis-specification testing) is the key to better testing. McCloskey, on the other hand, proposes reconsidering the ideal of testing and taking a look at other ways of arguing, like rhetorics. In my view, those attitudes are erroneous.

Let us start with the Popperian ideal. Many contributors to the empirical literature on testing homogeneity, perhaps ignorant of philosophy of science, pay lip-service to Popper. They are Popperian in their support of the rationality principle, but turn out to be non-Popperian in their stubborn refusal to give up demand theory. Should they be condemned, should they test better and harder? I do not think so. These able econometricians were not playing econometrics with the net down. Verification, not falsification, appears to be hard, and rewarding. It is difficult to reconcile the episode described in this chapter with a Popperian methodology. Both on the positive level, and on the normative level, the Popperian approach is wrong. Not surprisingly, Blaug (1980, chapter 6) is rather vague in his neo-Popperian interpretation of empirical research of consumer behaviour.

Neither is outright scepticism warranted. The homogeneity condition may not have been overturned, but many suggestions (made by economic theorists) as to how to measure elasticities have been scrutinized by econometricians. Some of the suggestions survived, others foundered. The literature shows fruitful cross fertilization: theory stimulated measurement, 'strange' test results stimulated theory (in particular, by inspiring new functional forms or a new meaning of particular test statistics). Hamminga (1983, p. 100) suggests that theory development and econometric research are independent from each other. The studies in consumer demand do not confirm this thesis.

What has been the significance of testing the null hypothesis of homogeneity? A rejection may not affect the null, but instead it points to the maintained hypothesis. A specification search is the result. A declaration of the falsification of the null of homogeneity would have been too simple and easy a conclusion. The reason why it is worthwhile to attempt to verify the 'meaningful theorems' is that the measurement of variables of interest, in particular elasticities, depends on the coherency of the measurement system, the demand equations. Phlips holds that if such measurement is to be meaningful, the restrictions of demand theory should be imposed. 'We find it difficult to take the results of these tests very seriously', Phlips (1974, p. 55) argues. If, however, the test is interpreted as a specification test, this difficulty disappears.

The research on homogeneity leads to the conclusion that the 'small world' of much empirical inference was not completely satisfactory. The empirical results did not lead to mechanistic updating of the probability of a hypothesis of homogeneity without scrutinizing auxiliary hypotheses. The homogeneity condition survived thanks to its strong prior support.

Notes

1. The two conditions are derived from revealed preference theory, from which the homogeneity condition can be deduced (Samuelson, [1947] 1983, pp. 111, 116). Compare with his remark (p. 97): 'As has been reiterated again and again, the utility analysis is meaningful only to the extent that it places hypothetical restrictions upon these demand functions.'
2. This chapter is partly based on Keuzenkamp (1994) and Keuzenkamp and Barten (1995).
3. Marschak's paper inspired Popper's 'zero method', the method of constructing a model on the assumption of complete rationality. See Popper ([1957] 1961, p. 141).
4. Cross (1982) suggests dropping the distinction between hard core and protective belt. This does not improve the transparency of Lakatos' methodology, however.

9 Positivism and the aims of econometrics

> Science for the past is a description, for the future a belief; it is not, and has never been, an explanation, if by this word is meant that science shows the necessity of any sequence of perceptions.
>
> Karl Pearson ([1892] 1911, p. 113)

1 Introduction

A standard characterization of econometrics is to 'put empirical flesh and blood on theoretical structures' (Johnston, [1963] 1984, p. 5), or 'the empirical determination of economic laws' (Theil, 1971, p. 1). For Chow (1983, p. 1) economic theory is the primary input of a statistical model. Since the days of Haavelmo (1944), econometrics almost serves as an afterthought in induction: the inductive stage itself is left to economic theory, the statistical work is about the 'missing numbers' of a preconceived theory.

This is not how the founders of probability theory thought about statistical inference. As Keynes ([1921] CW VIII, p. 359) writes,

> The first function of the theory is purely descriptive . . . The second function of the theory is inductive. It seeks to extend its description of certain characteristics of observed events to the corresponding characteristics of other events which have not been observed.

The question is to what extent econometrics can possibly serve those functions. In order to answer this question, I will characterize the aims of econometrics and evaluate whether they are achievable. I will argue that the only useful interpretation of econometrics is a positivist one. Section 2 deals with the philosophical maxims of positivism. In the subsequent sections, I discuss measurement (section 3), causal inference (section 4), prediction (section 5) and testing (section 6). A summary is given in section 7.

2 Positivism and realism

2.1 Maxims of positivism

What are 'the' aims of science? There is no agreement on this issue. It is useful to distinguish two views: positivism (related to empiricism, pragmatism and instrumentalism) on the one hand, and realism (which assumes the existence of 'true' knowledge, as in Popper's falsificationism) on the other hand. The following maxims (taken from Hacking, 1983, p. 41) characterize positivism:

(i) Emphasis on verification, confirmation or falsification
(ii) Pro observation
(iii) Anti cause
(iv) Downplaying explanations
(v) Anti theoretical entities.

The first relates to the *Wiener Kreis* criterion of meaningfulness, but also to the Popperian demarcation criterion. Popper does not support the other positivist maxims and, therefore, does not qualify as a positivist (p. 43). The second maxim summarizes the view that sensations and experience generate empirical knowledge or rational belief. The third is Hume's, who argues that causality cannot be inferred. The fourth maxim again is related to Humean scepticism on causality. Newton's laws, so it is argued, are not explanations but rather expressions of the regular occurrence of some phenomena (is gravitation a helpful though metaphysical notion, or is it a real causal force?). The fifth point deals with the debate between realists and anti-realists. I will make a detour on realism in the next section, but it may be helpful to state my own beliefs first.

De Finetti's one-liner 'PROBABILITY DOES NOT EXIST' (see chapter 4) and my counterpart 'the Data Generation Process does not exist' (see chapter 7, section 4.2), are both implicit endorsements of the positivist maxims. Interpreting the search for homogeneity as a specification search, serving the measurement of elasticities and approximation of demand equations, instead of a search for 'truth', is a positivist interpretation (chapter 8). Indeed, I claim that econometrics belongs to the positivist tradition. I am not at all convinced by Caldwell (1982), who argues that we are 'beyond positivism', and even less by post-modernists, like McCloskey (1985) who hold that it is all rhetorics. With Bentham (1789, p. 2), we should say: 'But enough of metaphor and declamation: it is not by such means that moral science is to be improved.'

I share the positivist's interest in confirmation and falsification (which is not the same as methodological falsificationism, which transforms common sense into dogma) and reject the rationalism or Cartesian doubt of someone like Frank Hahn. Hahn is sceptical about econometric confirmation and relies on his ability to think. His *cogito* cannot be better expressed than by Hahn himself: 'It is not a question of methodology whether Fisher or Hahn is right. It is plain that Hahn is' (1992, p. 5).

I also support the emphasis on measurement and observation. In sections 3.2 and 5.1 below, I will criticize economists who think that observation is unimportant. With respect to causality, I hold (like Hendry, in Hendry, Leamer and Poirier, 1990, p. 184) that econometrics should be Humean: we observe conjunctions, we do not 'know' their deeper mechanism. Still, it may be convenient to speak of causes, certainly if we have specific interventions in mind (this is natural for economists who are interested in policy). I interpret the word 'cause' as 'the rational expectation that such and such will happen after a specific intervention'. Finally, I believe explanations serve a purpose: they are useful in constructing analogies which may improve theoretical understanding. But I do not view understanding and explanation as roads to truth: they are psychological tools, useful in communicating policy analysis or intervention. Econometric models aim at empirical adequacy and useful interpretations, not truth. As truth does not have an objective existence in economics, I think it does not make sense to take scientific realism as the philosophical starting point of economic inference. Realism is not my specific interest, but I will discuss it briefly now.

2.2 *Scientific realism*

Realism takes the notions in science as 'real', existing objectively as the 'truth' which science should 'discover'. Hacking (1983) notes that there are two versions of realism: realism about theories (i.e. theories aim at the truth) and realism about entities (i.e. the objects described in theories really exist). Positivists do not care about the realism of either.

Feynman (1985) illustrates the problem of scientific realism by asking whether the inside of a stone exists. No one has ever seen such an inside; anyone who tries, by breaking the stone into two parts, only sees new outsides. Another example is an electron: do electrons exists or are they just models, inventions of the mind? Concepts like 'electron' may be useful to clarify our thoughts and to manipulate events, but that does not mean they are to be taken as literally true descriptions of real entities. They do not refer to a true state of nature, or true facts. If different physicists mean different things when they speak of an electron, then this electron is merely an image, an invention of the mind. This is reflected in van Fraassen's (1980, p. 214) question: 'Whose electron did Milikan observe; Lorentz's, Rutherford's, Bohr's or Schrödinger's?'

Economic counterparts of Milikan's electrons are not hard to think of. Money is an example. In the form of a penny or a dime money is pretty real, but its true value may be an illusion! And as an aggregate, like M1, it is highly problematic. Aggregation and index number theory suggest that M1 is a poor proxy for the money stock, and not a real entity corresponding to the theoretical entity 'money' occurring in a phrase like 'money causes income'. The money aggregate is a theoretical notion, a convenient fiction, which has no real existence outside the context of a specific theory.

Lawson, who supports realism, argues that realists hold that the object of research exists independently of the inquiry of which it is an object. 'True' theories can be obtained, he argues, and the objective world does exist (Lawson, 1987, p. 951). Lawson (1989), however, observes that econometricians tend to be instrumentalists. Instead of wondering what might be wrong with realism, he concludes that, because of their instrumentalist bias, econometricians have never quite been able to counter Keynes' critique on econometrics (see chapter 4, section 2.4). But I fear that a realist econometrics would make econometrics incomprehensible. Lawson does not clarify how 'realist econometrics' should look. Would it search for deep parameters (the 'true' constants of economics, usually those of consumption and production functions), in the spirit of Koopmans (1947) or Hansen and Singleton (1983)? Summers (1991) criticizes this approach: it yields a 'scientific illusion'.

Moreover, in economics, the object is not independent of the subject. In other words, behaviour (like option pricing) is affected by economics (the option pricing model). An independent truth does not exist. This also relates to the question of whether we *invent* or *discover* theories. Discovery belongs to realism. In economics, inventions matter.[1] The difference is that inventions are creations of our minds, not external realities. The price elasticity of demand for doughnuts is not a real entity, existing outside the context of a specific economic model. I concur with van Fraassen (1980, p. 5), who argues that

> scientific activity is one of construction rather than discovery: construction of models that must be adequate to the phenomena, and not discovery of truth concerning the unobservable.

A problem is how to define 'adequate'. Van Fraassen does not tell. I think there is at least a heuristic guide to determine relative adequacy: the Minimum Description Length principle (see chapter 5). It combines a number of the aims of inference, in particular descriptive and predictive performance. The MDL principle can also be invoked by those who emphasize other aims, such as finding causal relations, or improving understanding. This is because both aims can be viewed as derivatives of descriptive and predictive performance.

Finally, I will make some very brief remarks on the 'realism of assumptions'. The debate on realism raged in economics during the late fifties and early sixties, but the issues then were not quite the same as those discussed above. In 1953, Milton Friedman published a provoking *Essay on Positive Economics*, in which he denies the relevance of realistic assumptions in economic theory. What counts, according to Friedman, is the predictive quality of a theory. Frazer and Boland (1983) reinterpret Friedman's position as an instrumental kind of Popperianism. This seems a contradiction in terms, given Popper's disapproval of instrumentalism. Friedman's position is a clearly positivist one, the title of his work is well chosen. Although Friedman recommends testing theories (by means of their predictions), this does not make him a Popperian.[2]

3 Science is measurement

3.1 From measurement to statistics

The best expression of the view that science is measurement is Kelvin's dictum: 'When you can measure what you are speaking about and express it in numbers, you know something about it, but when you cannot measure it, you cannot express it in numbers, your knowledge is of a meagre and unsatisfactory kind.' Francis Galton and Karl Pearson advocate this principle in the domain of biometrics and eugenics. Galton even wanted to measure boredom, beauty and intelligence. It definitely is a nineteenth-century principle, but its popularity extends to the twentieth century including econometrics. The original motto of the Cowles Commission leaves no doubts: 'science is measurement'.[3] Measurement belongs to maxims of positivism in its emphasis on observation, although few positivists would argue that measurement is the ultimate goal of science.

The quest for measurement started with the search for 'natural constants'. The idea of natural constants emerged in the nineteenth century (when, for example, experimental physicists started to measure Newton's constant of gravitation; see Hacking, 1990, p. 55). Nineteenth-century social scientists tried to find 'constants of society', such as birth rates and budget spending. For example, Charles Babbage made a heroic effort to list all relevant constants that should be measured (pp. 59–60). Babbage and the Belgian statistician, Quetelet, are good examples of those who endorsed measurement for the sake of measurement.[4] If Babbage had known Marshall's concept of an elasticity, he would have added a large number of elasticities to his list. The first estimated consumer demand equations, due to Moore and Schultz (see chapter 8), are the economists' extensions of the positivist research programme. It is no surprise that Moore was a student of Karl Pearson, and Schultz was a pupil of Moore.

The rise of statistics parallels the growing availability of 'data'. This is one of the themes of Hacking (1990). The more numerical data that became available, the higher the demand for their reduction. As related in chapter 6, Galton was not only obsessed by measurement, but also invented the theory of regression and correlation. His ratings of beauty are based on the 'law of error' (normal law or Gaussian distribution), and this law serves his analysis of correlation. By making the steps from measurement, via the law of error, to correlation, Galton 'tamed chance' (Hacking, 1990, p. 186). Galton bridged description and understanding by means of statistics: correlation yields explanation, and the law of error is instrumental in the explanation. Correlation can be interpreted as a replacement for causes, causation is the conceptual limit to correlation.

3.2 Robbins contra *the statisticians*

During the founding years of econometrics, Lionel Robbins became a spokesman against econometrics. Unlike Keynes, he is not one of the founding (1933) fellows of the Econometric Society. Robbins, ([1932] 1935, p. 104) views economics as 'deductions from simple assumptions reflecting very elementary facts of general experience'. The validity of the deductions does not have to be established by empirical or inductive inference, but is

> known to us by immediate acquaintance... There is much less reason to doubt the counterpart in reality of the assumption of individual preferences than that of the assumption of an electron. (p. 105)

Hence, Robbins is a realist.[5] Robbins combines realism with an aversion to econometric inference. He rejects quantitative laws of demand and supply. If this were the result of academic conservatism, Robbins could be ignored. However, he makes a point that can be heard today as well. Among contemporary contributors to economic methodology, at least one (Caldwell, 1982) thinks that Robbins (and Austrian economists, who share many of his ideas) has a defensible case. Therefore, consider Robbins' argument. The problem with obtaining quantitative knowledge of economic relations, Robbins argues, is that it is plain that we are here entering upon a field of investigation where there is no reason to suppose that uniformities are to be discovered. The 'causes' which bring it about that the ultimate valuations prevailing at any moment are what they are, are heterogeneous in nature: there is no ground for supposing that the resultant effects should exhibit significant uniformity over time and space. No doubt there is a sense in which it can be argued that every random sample of the universe is the result of determinate causes. But there is no reason to suppose that the study of a random sample of random samples is likely to yield generalizations of any significance. (p. 107)

Clearly, this view is at odds with the view of Quetelet and later statisticians, who found that many empirical phenomena do behave uniformly, according to the 'law of error'. The central limit theorem (or its – invalid – inversion, see chapter 6, section 2) is of special importance in the investigations of those statisticians. In order to be persuasive, Robbins should have shown why, if there is no reason to expect stable relations, the early statistical researchers of social phenomena found so many regularities. Robbins could also have pointed out which conditions for the validity of the central limit theorem are violated in economics, or how it could be used in economics. Such arguments are not given. He just writes that it is 'plain'. (Remember Hahn!) The investigations of early econometricians on autonomous relations, identifications, etc., have been more useful to the understanding of possible weaknesses of econometric inferences, than Robbins' unsupported claims that there are simply no uniformities to be discovered.

Robbins attacks the idea that econometricians can derive quantitative statistical laws of economics. In a sense, Robbins is Pyrrhonian. His solution is the *a priori* strategy, which (he thinks) enables one to derive qualitative laws of economics.[6] Robbins ([1932] 1935, pp. 112–13) not only attacks the statistical analysis of demand and supply, but also statistical macroeconomics, in particular Wesley Mitchell's (1913) *Business Cycles*. Mitchell investigates the similarities between different business cycles, whereas Robbins argues that the only significance in describing different business cycles is to show their differences due to varying conditions over space and time. 'Realistic studies' (i.e. statistical investigations) may be useful in suggesting problems to be solved, but 'it is theory and theory alone which is capable of supplying the solution' (Robbins, [1932] 1935, p. 120).

Robbins' views are related to the Austrian methodology (see section 5.2). Most Austrians reject the predictive aim of economics, Hayek is an exception. He argues that statistics is useful only insofar as it yields forecasts. Robbins, on the other hand, claims that forecasts do not result from (unstable) quantitative statistical relations, but from (qualitative) economic laws (such as the law of diminishing marginal utility). The latter are 'on the same footing as other scientific laws' (p. 121). Exact quantitative predictions cannot be made. This is his reason for rejecting the search for statistical relations or 'laws'. Only if econometricians were able to estimate all elasticities of demand and supply, and if we could assume that they are natural constants, Robbins (pp. 131–2) claims, we might indeed conceive of a grand calculation which would enable an economic Laplace to foretell the economic appearance of our universe at any moment in the future. But such natural constants are outside the realm of economics. Robbins' view is not unlike Keynes' in his debate with Tinbergen. Keynes, though, is much closer to Hume's strategy of conventionalism and naturalism than to the neo-Kantian apriorism of Robbins. Also, Keynes' objections to the inductive aims of econometrics are better founded than those of Robbins. Unlike Keynes, Robbins creates a straw man, an economic Laplace to foretell the future. Few modern econometricians aim at this and neither did the founders. Moore, Schultz and other early econometricians followed Pearson, the positivist. He did not want to catch Laplace's demon in a statistical net. Pearson opposed necessity and determinism. His conception of (statistical) law is neither Cartesian, nor Laplace's:

> law in the scientific sense only describes in mental shorthand the sequences of our perceptions. It does not explain why those perceptions have a certain order, nor why that order repeats itself; the law discovered by science introduces no element of necessity into the sequence of our sense-impressions; it merely gives a concise statement of how changes are taking place. (Pearson, [1892] 1911, p. 113; see also pp. 86–7)

3.3 Facts and theory

The pioneer of the econometrics of consumer behaviour, Schultz, argues that in the 'slippery field of statistical economics, we must seek the support of both theory and observation' (1937, pp. 54–5). This is sound methodological advice. It neither gives theory nor observation a dominant weight in scientific inference, unlike the authors of the textbooks cited at the beginning of this chapter. Following Koopmans and Haavelmo, they put theory first.

Koopmans holds that fact finding is a waste of time if it is not guided by theory–neoclassical theory, that is. But fact finding is an important stage in science. This is acknowledged by Haavelmo (1944, p. 12; see also chapter 6), who speaks of a useful stage of 'cold-blooded empiricism'. Take Charles Darwin, who writes that he 'worked on true Baconian principles, and, without any theory, collected facts on a wholesale scale' (*Life and Letters of Charles Darwin*, cited – with admiration – in Pearson, [1892] 1911, p. 32). Darwin even tried to suppress his theory: 'I was so anxious to avoid prejudice, that I determined not for some time to write even the briefest sketch of it' (p. 33). Mitchell's (1913) method of business cycle research is like Darwin's, but Mitchell was not able to make the consecutive step: to invent a theory to classify the facts. This step has been made by followers of Mitchell, in particular, Kuznets, Friedman and Lucas. Lucas (1981, p. 16) praises the empirical work of Mitchell and Friedman for providing the 'facts' or evidence that a theory must cope with. It may surprise some readers to find Lucas, best known as a macro-economic theorist, in the company of Mitchell instead of Koopmans, but he rightly places himself in the positivist tradition.[7]

Stylized facts are a source of inspiration for economic theory. Kuznets' measurements of savings in the USA inspired Friedman to his permanent income theory of consumption. In particular, Kuznets found that the savings rate is fairly stable and does not (as Keynes suggested) decline with an increase in income. This is a good example of how measurement can yield a fruitful stylized fact and inspire new theory and measurements. Similarly, Solow's stylized facts about growth have inspired an immense literature on the basic characteristics and determinants of growth. There wouldn't have been an endogenous growth literature without studies such as Solow's. The example illustrates the way science proceeds: from simple (stylized fact) to more general (neoclassical growth models, endogenous growth models). It would be wrong to reject studies like Mitchell's, Friedman and Schwartz's, or Solow's because their stylized facts are all gross mis-specifications of economic reality.

But theory is an essential ingredient to classify facts. An obvious example is consumer behaviour. To qualify as a demand equation, an inferred regularity has to satisfy certain theoretical restrictions. Only then are the estimated elasticities credible. In the slippery field of empirical economics, we do indeed need both theory and observation.

3.4 Constants and testing

The quest for constants of nature and society started with Charles Babbage. Measuring nature's constants not only inspired positivist research, but is of specific interest for (neo-) Popperians, such as Klant. Without constants, Klant (1979, pp. 224–5) argues, the falsifiability principle for the demarcation of science breaks down. Popper ([1957] 1961) acknowledges that lack of constancy undermines methodological falsificationism. In economics, there is little reason to assume that parameters are really constant. Human beings change, as do their interactions and regularities. Logical falsificationism, therefore, is beyond reach. In principle, positivists do not have to be bothered by this problem. Their aim is not to discover and measure 'true' constants to single out scientific theories, but to invent measurement systems in order to measure parameters which are useful in specific contexts.

Charles Peirce contributed to this view. He denies the existence of natural constants. There are no Laws of Nature, hidden in the Book of God, with 'true' natural constants to be discovered. Instead, laws evolve from chance, their parameters are at best limiting values that will be reached in an indefinite future (see Hacking, 1990, p. 214). Perhaps this is even too optimistic about convergence to limiting values (see for example Mirowski, 1995, who discusses the 'bandwagon effect' of measurements of parameters in physics). The claim that physicists can apply the Popperian methodology of falsification, while economic theories are not strictly refutable, cannot be accepted. In both cases, strict methodological falsificationism is unfeasible. Econometricians face an additional problem: new knowledge, such as the publication of an empirical econometric model, may lead to a 'self-defeating excercise in economic history' when the findings are used to change the object of inference (Bowden, 1989, p. 258). This is the problem of reflexivity.

Because of the absense of universal constants, Klant argues, econometricians cannot test (i.e. falsify) general theories. Instead, they estimate specific models. But there are other kinds of testing, more subtle and much more relevant to econometrics than falsification (see section 6). Lack of experimentation, weakly informative data sets and the problem of reflexivity are more fundamental to econometrics than the presumed absence of universal natural constants. Reflexivity and other sources of changing behaviour may lead to time varying or unstable parameters, but this does not prohibit making predictions or learning from experience. Positivism does not leave the econometrician empty handed.

4 Causality, determinism and probabilistic inference

'A main task of economic theory is to provide causal hypotheses that can be confronted with data' (Aigner and Zellner, 1988, p. 1). 'Cause' is a popular word in econometrics, but whether causal inference really is a main or even the *ultimate* goal of econometrics may be questioned. The debate between rationalism (associated with scientific realism) and empiricism (or positivism) is to a large extent a debate on causality. Hence, we may expect at least two separate views on causation. Indeed, the empiricist holds that causation is a metaphysical notion that is not needed for scientific investigation. Science is about regularities, and that suffices. If you want to speak of causes, you may do so, but we cannot obtain true knowledge about the causes that go on behind the regularities observed. This view has been opposed by Kant, who argues that causation is the *conditio sine qua non* for scientific inference.[8] Without cause, there is no determinant of an effect, no regularity, no scientific law. The view dates back to the initiator of scepticism, Sextus Empiricus, who argues that 'if cause were non-existent everything would have been produced by everything and at random' (*Outlines of Pyrrhonism*, cited in van Fraassen, 1989, p. 97). The difference is that Kant provides an optimistic alternative to scepticism, by means of his *a priori* synthetic truth. Science presupposes the causal principle.

An intermediate position, which is positivist in content but pragmatic in form, is expressed by Zellner. He uses words such as 'cause' and 'explain' with great regularity and sympathy. But 'explain' means to him: fits well to past phenomena over a broad range of conditions (Zellner, 1988). 'Cause' is defined in terms of predictability: Zellner advocates Feigl's definition of 'predictability according to a law or set of laws' (this definition takes account only of sufficient causation; see below). A law as defined by Zellner, is a regularity that both 'explains' (i.e. fits with past phenomena) and predicts well. He adds a probabilistic consideration: models are causal with high probability when they perform well in explanation and prediction over a wide range of cases. Zellner's work is in the positivist tradition; his favourite example of a fruitful search for a causal law is Friedman's study of consumption. Zellner's approach to causality is sensible, although it has to be refined. Causal inference is primarily of interest if one has a specific intervention in mind, in which case 'cause' can be interpreted as 'the rational expectation that such and such will happen'. The type of intervention has to be made explicit, as well as the other conditions for the rational expectation.[9]

Before discussing causation in econometrics, I will deal with the meaning of causation in a deterministic world. Section 4.1.1 introduces Laplace's demon, the deterministic representative of the founder of probability theory. In section 4.1.2, necessary and sufficient causation is discussed, and the subjunctive conditional interpretation of causation is presented. Section 4.1.3 deals with Hume's critique on causal inference, section 4.1.4 discusses Peirce on determinism. In section 4.2, the relation between causation and probability is discussed. Section 4.3 extends the discussion to the domain of econometrics.

4.1 Causation and determinism

4.1.1 Laplace's demon

It is an 'evident principle that a thing cannot begin to be without a cause producing it', writes Laplace (cited in Krüger, 1987, p. 63). Laplace, a founder of probability theory, is a determinist. He subscribes to Kant's principle of universal causation (see below). All events follow the laws of nature. If one only knew those laws plus initial conditions, one could know the future without uncertainty. Please let me introduce Laplace's demon:

> We ought then to regard the present state of the universe as the effect of its previous state, and as the cause of that which will follow. An intelligence which for a given instant knew all the forces by which nature is animated, and the respective situations of the existences which compose it; if further that intelligence were vast enough to submit these given quantities to analysis; it would embrace in the same formula the greatest body in the universe and the lightest atom: nothing would be uncertain to it and the future as the past would be present to its eyes. (Laplace, 1814, p. vi, translation by Pearson, 1978, p. 656)

Nothing in the future is uncertain for the demon. The human mind searches for truth, Laplace continues, by applying the same methods as this vast intelligence. This makes men superior to animals. 'All men's efforts in the search for truth tend to carry him without halt towards such an intelligence as we have conceived, but from which he will always remain infinitely remote.' The reason why we need probability theory is that we are less able than the demon. There is an incomprehensible number of independent causal factors; probabilistic inference can help to obtain knowledge about the probability of causes. Chance is the reflection of ignorance with regard to the necessary or true causes.

4.1.2 Necessary and sufficient causes

What constitutes a 'cause'? Below, I discuss a number of definitions. A well known one is Galileo's efficient cause C of an event E, which is the necessary and sufficient condition for its appearing (Bunge, [1959] 1979, pp. 4, 33) if the following condition holds:

if and only if C, then E.

Inferring the efficient cause of event E is straightforward, for it implies that not-C results in not-E. A problem is that many events are determined by a very large or even infinite amount of other events. This definition is, therefore, of little use. A weakened version is the sufficient but not necessary causal condition, if C, then E.

This is how John Stuart Mill defines causes, with the additional provision that the cause precedes the event in time.[10] The definition facilitates testing for causality. The principle of the uniformity of nature, stating that the same cause must always have the same effect, supplements this causal relation with immutable laws of nature.

These are the most elementary notions of cause. But how can we know that causes are real, rather than spurious? One way might be to distrust our sensations, the phenomena. Rational reasoning, deduction, will reveal true causes. This view is exemplified by Descartes who holds that '[n]othing exists of which it is not possible to ask what is the cause why it exists' (cited in Bunge, [1959] 1979, p. 229, n. 10). This proposition is better known as Leibniz's *principle of sufficient reason*, according to which nothing happens without a sufficient reason or cause (Leibniz uses reason and cause as synonyms; for Descartes and Leibniz, reasoning and obtaining causal knowledge were nearly identical). But this principle is of little help to demarcate spurious and real causes. Hume's scepticism is a critique of Descartes' rationalistic view on causation.

Kant tried to reconcile Descartes and Hume, by making a distinction between 'das Ding an sich', i.e. the 'thing in itself', and its appearance, 'das Ding für sich'. Although Kant accepted that we cannot obtain knowledge of things in themselves, we are able to make inferences with respect to their appearances. The principle of universal causation states that appearances of things, i.e. phenomena, are always caused by other phenomena: 'Alles zufällig Existierende hat eine Ursache' (everything which happens to exist has a cause; cited in Krüger, 1987, p. 84, n. 10; see also von Mises, [1928] 1981, p. 210).[11] Kant's principle is introduced for the purpose of making inference possible and has an epistemological intention. It relates to experiences, not to 'das Ding an sich'. But the principle is not empirically testable, it is an *a priori* synthetic proposition, a precondition for empirical research (see also Bunge, [1959] 1979, pp. 27–8). Hume would have considered this requirement superfluous.

An alternative to the definitions of causations presented above is to base the notion of cause on the so-called subjunctive conditional. This is a proposition of what would happen in a hypothetical situation, i.e. if A had been the case, B would be the case. According to the subjunctive conditional definition of cause, C is a cause of E if the following condition is satisfied: if C had not occurred, E would not have occurred. This is anticipated in Hume ([1748] 1977, p. 51): 'we may define a cause to be an object, followed by another, and where all the objects, similar to the first, are followed by objects similar to the second. Or in other words, where, if the first object had not been, the second never had existed' (n.b., this final sentence is not correct).

4.1.3 Hume's critique of causality

According to Hume, you may call some things causes, if it pleases you, but you cannot *know* they are. Causation is convenient shorthand for constant conjunction. Reasoning will not enable one to obtain causal knowledge. Hume ([1739] 1962) argues that there are three ingredients in apparently causal relations:

(i) contiguity (closeness) in space and time
(ii) succession in time
(iii) constant conjunction: whenever C, then E, i.e. the same relation holds universally.

The Cartesian theory of causality holds that everything must have a sufficient cause, and some events must be or cannot be the causes of other events. But, Hume argues, true knowledge of necessary causes cannot be derived from observation, whence the problem of induction and Hume's scepticism. At best, we observe relations of constant conjunction. If they also obey (i) and (ii), by way of habit we call them causal laws. They are subjective mental constructs, psychological anticipations, not truth statements of which the validity can be ascertained. Bunge ([1959] 1979, p. 46), a modern realist, criticizes Hume:

> the reduction of causation to regular association, as proposed by Humeans, amounts to mistaking causation for one of its tests; and such a reduction of an ontological category to a methodological criterion is the consequence of epistemological tenets of empiricism, rather than a result of an unprejudiced analysis of the laws of nature.

Bunge is correct that Humean causation is indeed of an epistemological nature, but this is not a mistake of Hume's: he explicitly denies that ontological knowledge of causation is possible. Bunge's own approach, summarized by the proposition that causation is a particular case of production (p. 47),[12] suffers from the same problems as the ones which Hume discusses: even if we experiment, we cannot know the true causes, as there is always a possibility that the observed regularities are spurious.

Habit, which results from experience, is Hume's ultimate justification for inference. If we regularly experience a constant conjunction of heat and flame, we may expect to observe such a conjunction in a next instance (Hume, [1748] 1977, p. 28). Without the guidance of custom, there would be an end to all human action. This view is shared by Keynes. Custom is the very guide in life, Hume argues in his naturalistic answer to scepticism.

There are no probabilistic considerations in Hume's discussion of causality (he deals with deterministic laws). Due to the influence of Newton's *Principia* (published in 1687), until the end of the nineteenth century 'explanation' meant virtually the same as 'causal description' (von Mises, [1928] 1981, p. 204). Gravitation figures as the most prominent 'cause' in Newton's laws of mechanics.[13] It is no surprise that, after the quantum revolution in physics, not only Newtonian mechanics came under fire but also this association of explanation and causality. In a famous essay, Russell (1913) argues that it is better to abandon the notion of causality altogether. According to Russell, the causality principle (same cause, same effect) was so popular because the idea of a function was unfamiliar to earlier philosophers. Constant scientific laws do not presume sameness of causes, but sameness of relations, more specifically, sameness of differential equations (as noted on p. 102 above, Russell claims in his *Outline of Philosophy* that scientific laws can only be expressed as differential equations; Russell, 1927, p. 122; see, for the same argument, Jeffreys, [1931] 1957, p. 189, who concludes: 'the principle of causality adds nothing useful').

There are certainly problems with the Humean view on causality. For example, the requirement of contiguity is not only questionable in physics (think of 'action at distance') but also economics (printing money in the US may cause inflation in the UK). Moreover, succession in time is not necessary for a number of relations that would be called causal by common sense (like 'force causes acceleration', Bunge, [1959] 1979, p. 63, or upward shift of demand curves causes higher prices in general equilibrium). Other definitions of causality are, therefore, needed if the notion of causality is to be retained. The alternative is more rewarding: abandoning the quest for true causes.

4.1.4 Peirce on the doctrine of necessity

Peirce (1892, p. 162) opposes Kant's principle of universal causation and other beliefs that 'every single fact in the universe is precisely determined by law'. Peirce exemplifies the changes of scientific thinking, from the deterministic nineteenth century, to the probabilistic twentieth century (see e.g. Hacking, 1990, pp. 200–3). Peirce provides the following arguments against the 'doctrine of necessity'.

First, inference is always experiential and provisional. A postulate of universal causation is like arguing that 'if a man should come to borrow money . . . when asked for his security . . . replies he "postulated" the loan' (Peirce, 1892, p. 164). Inference does not depend on such a postulate, science deals with experience, not with things in themselves. Here, Peirce joins Hume in his emphasis on observable regularities.

Secondly, Peirce denies that there are constant, objectively given parameters (see above). The necessitarian view depends on the assumption that such parameters do have fixed and exact values. Peirce also denies that exact relationships can be established by experimental methods: there always remain measurement errors (p. 169). Errors can be reduced by statistical methods (e.g. least squares), but 'an error indefinitely small is indefinitely improbable'. This may be seen as an invalid critique of the doctrine of necessity: the fact that we cannot measure without error the exact value of a continuous variable in itself does not undermine the possible existence of this value. But then, Peirce argues, a belief in such existence must be founded on something other than observation. In other words, it is a postulate – and such postulates are redundant in scientific inference.

The fact that we do observe regularities in nature does not imply that everything is governed by regularities, neither does it imply that the regularities are exact:

> Try to verify any law of nature, and you will find that the more precise your observations, the more certain they will be to show irregular departures from the law... Trace their causes back far enough, and you will be forced to admit they are always due to arbitrary determination, or chance. (Peirce, 1892, p. 170)

Peirce was writing before the quantum revolution in physics, but the probabilistic perspective made him reject Laplace's determinism. The fact that there still is regularity may be the result of chance events given the 'law of error', the normal distribution (as was argued by Quetelet and Galton). Regularity may be the result of laws, even of probabilistic laws. It can equally well result from evolution and habit. But diversity can never be the result of laws which are immutable, i.e. laws that presume that the intrinsic complexity of a system is given once and for all (see Peirce, 1892, p. 173). Chance drives the universe.

4.2 Cause and chance

4.2.1 Laplace revisited

Despite his determinism, Laplace links probability and causality. For him, there is no contradiction involved. Laplace regards probability as the result of incomplete knowledge. He distinguishes constant (regular) causes from dynamic (irregular) causes, 'the action of regular causes and constant causes ought to render them superior in the long run to the effects of irregular causes' (cited from the *Essai Philosophique sur les*

Probabilités, pp. LIII–LIV, translation Karl Pearson, 1978, pp. 653-4). Our ignorance of causes underlies probabilistic inference. As this ignorance is the greatest in the 'moral sciences', one might expect Laplace to recommend probabilistic inference especially to those branches. This, indeed, he does, in a section on application of the calculus of probabilities to the moral sciences.

Consider an urn, with a fixed but unknown proportion of white and black balls. The constant cause operating in sampling balls from the urn is this fixed proportion. The irregular causes of selecting a particular ball are those depending on movements of the hand, shaking of the urn, etc. Laplace's demon can calculate the effects of the latter, but human beings cannot. What can be done is to make probabilistic propositions about the chance of drawing particular combinations of balls. The irregular causes disappear, but a probability distribution remains. For a long period, this view on probability and causation influenced probability theory. For example, Antoine Augustin Cournot, of a younger generation of statisticians than Laplace, still sticks to the view that apparently random events can result from a number of independent causal chains (Hacking, 1975, p. 174; Krüger, 1987, p. 72). The change came at the end of the nineteenth century, when the deterministic world-view began to crumble. Meanwhile, the frequency interpretation of probability developed. How does it relate to causation?

4.2.2 Frequency and cause

Since Hume, one of the positivist maxims has been scepticism about causes. This is reflected in the writings of Ernst Mach, and his followers Karl Pearson and Richard von Mises. Mach holds that the notion of causation is an out-dated fetish that is gradually being replaced by functional laws (Bunge, [1959] 1979, p. 29). This is similar to Russell's belief that causal laws will be replaced by differential equations. Pearson goes a step further and claims that causal laws will be replaced by empirical statistical correlations. Pearson ([1892] 1911) can be read as an anti-causal manifesto. He is a radical Humean, arguing that cause is meaningless, apart from serving as a useful economy of thought (pp. 128, 170). Cause is a 'mental limit', based upon our experience of correlation and contingency (p. 153).

Von Mises' thinking on causality is more delicate. He is a positivist, inspired by Mach, and his probability theory owes some ideas to the posthumously published book *Kollektivmasslehre* (1897) of G. Theodor Fechner (see von Mises, [1928] 1981, p. 83). Both Mach and Fechner reject the principle of universal causation. Fechner also rejects determinism, largely for the same reasons as Peirce. Fechner's sources of indeterminism are inaccuracy, the suspension of causal laws, the occurrence of intrinsically novel initial conditions and, finally, the occurrence of self-fulfilling prophecies (reflexivity).

While Mach, Fechner and Pearson reject causalism out of hand, von Mises does not think that one needs to abandon the notion of causality in (probabilistic) inference. In discussing the 'small causes, large effects' doctrine of Poincaré, von Mises notes that statistical theories (such as Boltzman's gas theory) do not contradict the principle of causality. Using Galton's Board (the 'quincunx', see chapter 6, section 3.1) as illustration, von Mises argues that a statistical analysis of the resulting distribution of the balls is not in disagreement with a deterministic theory (although the latter would be very hard to implement to analysing the empirical distribution). The statistical theory does not compete with a deterministic one, but is another form of it (von Mises, [1928] 1981, p. 209). Von Mises even considers that the statistical distributions of events like throwing dice, or rolling balls on Galton's Board, may be given a causal interpretation. The meaning of the causal principle changes. Causality in the age of quantum physics is not quite the same as it was during the heyday of mechanical determinism.

Von Mises (p. 208) argues that classical mechanics is of little help in 'explaining' (by means of causal relations) semi-stochastic phenomena, such as the motion of a great number of uniform steel balls on Galton's Board. What is needed is a

> simple assumption from which all the observed phenomena of this kind can be derived. Then at last can we feel that we have given a causal explanation of the phenomena under investigation.

Hence, von Mises claims that causal explanation is directly related to the simplicity of assumptions. But cause is not an absolute notion, as it is in the writings of Descartes and his followers. A theory is not conditional on the existence of real causes (as it is in the views of Descartes and of Kant), but conversely: causes are conditional on (fallible) knowledge of theories based on empirical observation. Causality is relative to theories formulated at some date, without claiming universal validity. Therefore, von Mises (p. 211) argues, the principle of causality is subject to change, it depends on our cognition. I regard this as a tenable position in positivism.

4.2.3 Keynes' causa cognoscendi

Unlike Peirce, who was able to anticipate the full consequences of probabilism, Keynes is a determinist: 'none of the adherents of "objective chance" wish to question the determinist character of natural order'

([1921] CW VIII, p. 317); objective chance results from 'the coincidence of forces and circumstances so numerous and complex that knowledge sufficient for its prediction is of a kind altogether out of reach' (p. 326). Despite this support for Laplace's perspective, Keynes does not make universal causation the ultimate foundation of his theory of probability and induction. In this sense, he is a Humean. Still, in a brief exposition on causality, Keynes (pp. 306–8) makes a few remarks on causation that anticipate the modern theory of probabilistic causation of Suppes (see e.g. Vercelli and Dimitri, 1992, pp. 416–17). Keynes distinguishes *causa essendi* (the ontological cause, where necessary and sufficient conditions can be stated) from *causa cognoscendi* (causality relative to other knowledge, which is a probabilistic notion dealing with regular conjunction).

Keynes provides a number of definitions for types of causes. There are two given propositions, e and c, related to events E and C where C occurs prior to E (remember that Keynes' theory of probability is formulated in terms of propositions). Furthermore, we have laws of nature (independent of time) l and other existential knowledge (facts), given by propositions f. The most elementary causal relations (that do not rely on f) are the following.

(i) If $P(e|c, l) = 1$, then C is a sufficient cause of E

(ii) If $P(e|\text{not-}c, l) = 0$, then C is a necessary cause of E.

The principle of universal causation (or law of causation, as Keynes calls it) is that, if l includes all laws of nature, and if e is true, there is always another true proposition c, such that $P(e|c, l) = 1$. Hence, this principle is about sufficient causes, i.e. case (i). But for the practical problem of induction, Keynes (pp. 276, 306) argues, the laws of universal causation and the uniformity of nature (same cause, same effect) are of little interest. Science deals primarily, with 'possible' causes. As an example (not given by Keynes), one might think of smoking: neither a necessary, nor a sufficient, cause of lung cancer. In order to deal with possible causation, Keynes first weakens definitions (i) and (ii). Consider the sufficient cause with respect to background knowledge f (case (iv) in Keynes, [1921] CW VIII; I will skip the similarly weakened versions of necessary causes):

(iv) If $P(e|c, l, f) = 1$ and $P(e|l, f) \neq 1$ then C is sufficient cause of E under conditions f.

This introduces background knowledge and makes a more interesting example of causation than the (unconditional) case (i), although smoking still is not a sufficient cause (type (iv)) of cancer. The step to possible causation (Suppes, 1970, uses the term *prima facie* cause in this context) is made by introducing a further existential proposition h, such that

(viii) If $P(h|c, l, f) \neq 0$ (i.e. the additional proposition is not inconsistent with the possible cause, laws and other facts),

$P(e|h, l) \neq 1$ (i.e. the effect does not obtain with absolute certainty in absence of c) and

$P(e|c, h, l, f) = 1$ (i.e. the effect is 'true'),

then C is, relative to the laws l, a possible sufficient cause of E under conditions f.

A possible cause can also be defined for the necessary case, yielding the somewhat odd 'possible necessary' cause. An interesting feature of Keynes' analysis of causation is that he is one of the first to provide a probabilistic treatment of causation. The recent theory of Suppes (1970) is not much different from Keynes' short treatment. But, unlike Suppes, Keynes does not have much interest in the types of causes such as presented above. They relate to *causa essendi*, whereas observations only make inference with respect to *causa cognoscendi* possible. A definition of *causa cognoscendi* is given:

If $P(e|c, h, l, f) \neq P(e|\text{not-}c, h, l, f)$, then we have 'dependence for probability' of c and e, and c is causa cognoscendi for e, relative to data l and f.

Keynes' *causa cognoscendi* brings us back to Humean regularity, which in his probabilistic treatment is translated to statistical dependence (and correlation, in the context of the linear least squares model; see Swamy and von zur Muehlen, 1988, who emphasize that uncorrelatedness does not imply independence). This, and not *causa essendi*, is what matters in practice:

The theory of causality is only important because it is thought that by means of its assumptions light can be thrown by the experience of one phenomenon upon the expectation of another. (Keynes, [1921] CW VIII, p. 308)

Keynes is aware that correlation is not the same as causation (although Yule's famous treatment of spurious correlation appeared five years later, in 1926). It is not unlikely that Keynes would hold that thinking (intuitive logic) would be the ultimate safeguard against spurious causation.

4.2.4 From Fisher's dictum to the simplicity postulate

Fisher argues that his inductive methods are intended for reasoning from the sample to the population, from the consequence to the cause (Fisher, 1955, p. 69), or simply for inferring the 'probability of causes' (p. 106). There is a clear link between randomness and causal inference in Fisher's writings. He notes that there is a multiplicity of causes that operate in agricultural experiments. Few of them are of interest. Those few can be singled out by sophisticated experimental design. Randomization is the first aid to causal inference (see the discussion of randomization and of experiments in econometrics in chapters 6 and 7).

Cause is a regularly recurring word in Fisher's writings, but he does not subscribe to a principle of causation such as Kant's. He is an exponent of British empiricism, his views on causality are not much different from Hume's (who is not mentioned in Fisher's books). Cause is used as a convenient way of expression, not unlike Pearson's 'mental shorthand'. There is no suggestion in Fisher's writings that he supports either a deterministic or an indeterministic philosophy.

One of Fisher's statements with regard to causality is interesting. When Fisher was asked at a conference how one could make the step from association to causation, Fisher's answer was 'make your theories elaborate' (Cox, 1992, p. 292). This dictum can be interpreted in various ways. One is to control explicitly for nuisance causes, by including them as variables in statistical models. This is discussed in Fisher ([1935] 1966, Chapter IX) as the method of concomitant measurement.[14] It amounts to making implicit *ceteris paribus* clauses explicit. But modelling all those additional factors can be fraught with hazards because the specific functional relations and interactions may be very complex. The alternative is to improve the experimental design.

The discussion of Pratt and Schlaifer (1988) on the difference between (causal) laws and regressions elaborates Fisher's ideas on concomitant measurement. Pratt and Schlaifer (p. 44, italics in original) argue that a regression is only persuasive as a law, if one includes in the regression

> *every 'optional' concomitant ... that might reasonably be suspected of either affecting* or merely predicting *Y given X – or if the available degrees of freedom do not permit this, then in at least one of several equations fitted to the data.*

In macro-econometrics, degrees of freedom are usually too low to implement this approach. A combination with Leamer's (1978) sensitivity analysis may be helpful. Levine and Renelt (1992) make such an effort. In micro-econometrics, when large numbers of observations are available, the argument of Pratt and Schlaifer may clarify (implicitly) why investigators do not vary significance levels of *t*-statistics with sample size (the same 0.05 recurs in investigations with 30 and 3000 observations). Including more 'concomitants', where possible, makes it more likely that the inferences are like laws instead of being spurious.

The modified simplicity postulate suggests why a researcher should not include all available regressors in a regression equation. One may even argue that this postulate enables a researcher to ignore remote possibilities, and constrain oneself to relatively simple models unless there is specific reason to do otherwise. This, I think, is Jeffreys' (and, perhaps, even Fisher's own) approach. Jeffreys ([1939] 1961, p. 11) criticizes the principle of causality, or uniformity of nature, in its form '[p]recisely similar antecedents lead to precisely similar consequences'. A first objection of Jeffreys is that antecedents are never quite the same, '[i]f "precisely the same" is intended as a matter of absolute truth, we cannot achieve it' ([1939] 1961, p. 12). More interestingly, Jeffreys asks how we may know that antecedents are the same. Even in carefully controlled experiments, such knowledge cannot be obtained. The only thing that can be done, is to control for some conditions that seem to be relevant, and hope that neglected variables are irrelevant. The question then arises, Jeffreys asks, 'How do we know that the neglected variables are irrelevant? Only by actually allowing them to vary and verifying that there is no associated variation in the result' (p. 12). This verification needs a theory of 'significance tests' (Jeffreys' Bayesian approach, or in Fisher's case, the approach outlined in chapter 3, section 3). Analysing the residuals in regression equations is an application of this line of thought.

4.3 Causal inference and econometrics

4.3.1 Recursive systems

Not surprisingly, causality was much discussed during the founding years of econometrics, in particular with respect to business cycle research. Tinbergen (1927, p. 715) argues that the goal of correlation analysis is to find causal relations. But he warns of the fallacy of 'a certain statistician, who discovered a correlation between great fires and the use of fire engines, and wanted to abolish fire engines in order to prevent great fires' (my translation). The same warning appears in nearly all introductory textbooks on statistics. Apart from this remark, Tinbergen does not dig into the problem of causal inference. Koopmans is more specific. Koopmans (1941, p. 160) argues that the 'fundamental working hypothesis' of econometric business cycle research is that causal connections between the variables are dominant in determining the fluctuations of the internal variables, while, apart from external influences readily recognized but not easily expressible as emanating from certain measurable phenomena, mere chance fluctuations in the internal variables are secondary in quantitative importance.

Koopmans defines causal connections as necessary. He also argues that this working hypothesis is not contradicted by the data (suggesting that it is testable). There are many interrelations between economic variables, which 'leaves ample freedom for the construction of supposed causal connections between them ... In fact, it leaves too much freedom' (pp. 161–2). Here, the economist should provide additional, *a priori* information. The economist selects a number of possible causal relations. The econometrician uses the 'principle of statistical censorship' (p. 163) to purge those suggested relations that are in conflict with the data.

Koopmans' methodological views on causality do not correspond to the positivist views. Compare Pearson, who argues that causality is just mental shorthand, adding nothing of interest to the statistical information of contingency or correlation. Koopmans, on the other hand, takes causality as a necessary requirement for business cycle research (like Kant, who argued that scientific inference needs a causal principle). Koopmans is less dogmatic than Kant, for he claims that his causal principle is just a working hypothesis, not rejected by the data so far. Furthermore, Koopmans (a physicist who grew up in the quantum revolution) does not deny the possibility of pure chance phenomena.

The publications of the Cowles Commission that followed elaborate Koopmans' point of view. Causality is an *a priori* notion, imposed as a property of models. In this sense, Cowles-causality can be regarded as different from Kant's causality, where it is a property of the 'things' (data) 'in themselves', hence (if we consider this in the current context), a property of the outcome space. Cowles-causation stands half way between the positivist notion of causality and the *a priori* view.

Simon (1953) and Strotz and Wold (1960) have tried to elaborate the definition of causality in terms of properties of (econometric) models (hence, not in terms of real world events). Simon's definition is based on restrictions on an outcome space S, where there are (at least) two sets of restrictions, A and B (the following is a simplified version of the example given in Geweke, 1982). An econometric model is the conjunction of those restrictions, $A \cap B$. One may think of restrictions on the determination of money, M, and income, Y:

$$A : M = a, \tag{1}$$

$$B : Y + bM = c. \tag{2}$$

The outcome space of this example is $S = \{(M, Y) \in \mathbb{R}^2\}$. This is an example of a causal ordering: condition (1) only restricts M, and condition (2) restricts Y without further restricting M. More generally, Simon defines a causal ordering as follows. The ordered pair (A,B) of restrictions on S determines a causal ordering from M to Y if and only if the mapping $G_Y(A) = Y$ and the mapping $G_X(A \cap B) = G_X(A)$. Hence, the causal ordering is a property of the model, not the data.

A related interpretation of causality is the causal chain model of Herman Wold. This causal chain can be represented by triangular recursive stochastic systems (see e.g. Wold and Juréen, 1953). Consider the model

$$\mathrm{By} + \Gamma\mathrm{x} = \mathrm{u}. \tag{3}$$

where B is a triangular matrix (with unit elements on the diagonal), and $\Sigma = \mathrm{E}(\mathrm{uu}')$ a diagonal matrix. An example (taken from Strotz and Wold, 1960) which is the stochastic counterpart to (1)–(2) is:

$$x = u_1 \sim N(\mu_1, \sigma_1^2), \tag{4}$$

$$y + bx = u_2 \sim N(\mu_2, \sigma_2^2). \tag{5}$$

The reduced form equations of such a system can conveniently be estimated with ordinary least squares, which yields consistent estimates. The causal chain model has two motivations. The first is that reality is supposed to follow a causal chain: 'truth' is recursive (a popular method of the Scandinavian school of economics, sequence analysis, is based on this postulate; see Epstein, 1987, p. 156). Wold even claims that the only way to deal with causal models is via recursive systems (see the critique of Basmann, 1963). A second motivation is that recursive systems avoid cumbersome simultaneous equations estimation techniques (in particular, Full Information Maximum Likelihood). *Qua* method, Wold opposes the Cowles approach. *Qua* philosophy, they are alike: they are close to the Kantian *a priori* synthetic truth. The assumptions underlying Wold's arguments (triangular system, normality) may be invoked as useful simplifications, but they remain assumptions. The modified simplicity postulate suggests that Wold's argument may be useful where the resulting descriptions perform well (evaluated for example by the Minimum Description Length principle), but it is not justified to argue that economic reality is *per se* a triangular recursive system. Another objection to Wold's approach is that, if there are k endogenous variables in the model, the number of possible causal orderings is $k!$. Wold does not say how to choose among those alternative specifications.

Recursive systems underlie many early macro-econometric models, such as Tinbergen's.

4.3.2 Causality and conditional expectation

Pearson ([1892] 1911, p. 173) argues that the unity of science is in its method, and the methods of science are classification, measurement, statistical analysis:

> The aim of science ceases to be the discovery of 'cause' and 'effect'; in order to predict future experience it seeks out the phenomena which are most highly correlated...From this standpoint it finds no distinction in kind but only in degree between the data, method of treatment, or the resulting 'laws' of chemical, physical, biological or sociological investigations.

Pearson's statement has a weakness. He does not consider the possibility of spurious and nonsense correlation (although classification can be viewed as his way of discriminating sense from nonsense). Here, the paths of Pearson and his student, Yule, diverge. Yule is interested in regression, conditional expectation. This surpasses Pearson's emphasis on association (correlation). An important reason why econometricians are interested in causation is that they want to avoid being accused of inferring 'spurious correlations', such as Tinbergen's (1927) example, fire engines cause great fires. The fallacy in his example can be exposed by a few simple experiments: it is straightforward to demonstrate that fire engines are neither sufficient nor necessary causes of great fires. In economics, it is not easy to obtain such knowledge by an appropriate experiment.

The conditional expectation model, which underlies regression analysis, has been used to clarify the notion of causality in cases of non-experimental inference. An example is Suppes (1970, p. 12). He distinguishes *prima facie* causes and spurious causes. Basically the same distinction is made in Cox (1992), who provides slightly more simple definitions. Cox's definitions are as follows. C is a *candidate cause* (*prima facie cause*, in Suppes' terminology) of E if $P(E|C) > P(E|\text{not-}C)$. Hence, candidate causes result at least in positive association. Next, C is a spurious cause of E if B explains the association, i.e. $P(E|C, B) = P(E|\text{not-}C, B)$. Cox's and Suppes' approaches brings us back to Fisher's dictum: elaborate your model. As it may always be the case that there is a variable B that has not been considered for the model, statistical inferences generate knowledge about candidate causes, not real causes. This is consistent with Hume's scepticism about establishing knowledge for necessary causes. Most econometricians, I think, will agree with Cox's view.

The philosopher Nancy Cartwright (1989), however, has tried to rehabilitate 'true' causes or 'capacities'. The initial motto of the Cowles Commission, 'Science is Measurement', is her starting point. However, measurement is not her single aim, it is supplemented by the search for causal relations, or more generally, 'capacities'. She summarizes her views as 'Science is measurement; capacities can be measured; and science cannot be understood without them' (p. 1). Cartwright prefers the study of capacities to the study of laws. She follows J. S. Mill by opposing Hume's view that the notion of cause should be replaced by the notion of regularity and constant conjunction. Every correlation, Cartwright (p. 29) argues, has a causal explanation, and the statisticians' warning that correlation does not entail causation is too pessimistic. Her theory involves the full set of (INUS) causal conditions for event E (each of them may be a genuine cause). If each of the elements in this set has an 'open back path' with respect to E, then all of them are genuine causes (p. 33). C has an open back path with respect to E if and only if C has another (preceding) cause C', and if it is known to be true that C' can cause E only by causing C. The condition is needed to rule out spurious causes. This, Cartwright (pp. 34–5) argues, is how 'you can get from probabilities to causes after all'. The major weaknesses in the argument are the requirement for having a full set of INUS conditions, and the presumed true knowledge in the 'open back path' condition.

Cartwright claims that her philosophy applies to the social sciences. The methods of econometrics, she argues (p. 158), presuppose that the phenomena of economic life are governed by capacities.[15] I think she overstates her case. To see why, consider the investigations of consumer demand presented in chapter 8. Measurement was clearly an aim of most demand studies. But which 'capacity' or cause was at stake? Should we understand 'capacity' in a very general sense, e.g. as rational (intentional) behaviour, utility maximization? I do not think that it is helpful to maintain that an elasticity of 2 is 'caused' by utility maximization of a group of individuals. But perhaps the elasticity is to be interpreted as the bridge by which a cause operates, a 1% change in income 'causes' a 2% change in consumption. Indeed, an elasticity is a proposition of the kind *'if A, then B'*. But the econometric investigations of such elasticities are based on regular conjunction, causation is only of interest if one wants to go beyond measurement, in particular, if intervention is the aim. Even in that case, 'cause' is mental shorthand for rational expectation. We can do without metaphysical capacities.

4.3.3 Wiener-Granger causality

Granger (1969) defines causality in terms of information precedence (or predictability) for stationary processes. It is an attempt to provide a statistical, operational definition that can help one to escape from the philosophical muddle. Granger's definition resembles Cox's of a non-spurious candidate cause, although Granger restricts the definition to processes that have finite moments. Formally, let all information in the 'universe' at time t be given by $\mathbf{U}_t$, $\mathbf{U}_t - \mathbf{X}_t$ represents all information except that contained in $\mathbf{x}_t$ and its past. Then $\mathbf{x}_t$ is said to Granger-cause $\mathbf{y}_{t+1}$ if

$$P(\mathbf{y}_{t+1} \text{ in } C|\mathbf{U}_t) \neq \mathrm{P}(\mathbf{y}_{t+1} \text{ in } C|\mathbf{U}_t - \mathbf{X}_t), \tag{6}$$

for any region C.[16] The definition contains a non-operational information set $\mathbf{U}_t$. This requirement resembles Carnap's (1950) requirement of total evidence, by which Carnap tried to avoid the so-called reference class problem. In order to obtain a more useful notion of causality based on the idea of precedence, one usually invokes a background theory on which a restricted information set I is based. Granger suggests using the term *'prima facie cause'* for this case, and this condition indeed corresponds to Suppes' *prima facie cause* as well as Cox's candidate cause. They all are probabilistic versions of conditions (ii) and (iii) of Humean causation.

Granger's criterion has a number of weaknesses and many of them have been discussed by Granger himself. I have already mentioned the assumption of finite moments. For non-stationary processes, one has to filter the time series in order to make them stationary. There are many ways to do this. The causality test is sensitive to the choice of filtering. Furthermore, filtering data is likely to abolish relevant information and, therefore, violates Fisher's reductionist approach to inference as well as inference based on the modified simplicity postulate.

There are other weaknesses. According to Granger's definition, sunspots 'Granger-cause' the business cycle (Sheenan and Grives, 1982). Sims (1972, p. 543) writes that the Granger causality test 'does rest on a sophisticated version of the *post hoc ergo propter hoc* principle'. Whether the characterization 'sophisticated' is right, depends on one's assessment of the importance of the choice of background information, I, in specific applications. This choice is crucial. Different choices of I may give different causality results (as Sims' sequence of tests of money–income causality, starting with Sims, 1972, shows). Granger's test for causality is highly sensitive to specification uncertainty. The approach is one of overreaching instrumentalism, where only predictive performance matters. While the Wold approach to causal inference exaggerates *a priori* economic theory, 'Granger-causality' neglects considerations of economic theory.

4.3.4 Cause and intervention

The notions of sufficient and necessary cause serve economic policy. If a policy maker knew the causes of recessions, he would like to neutralize the necessary cause of it, or create a sufficient cause for a boom. But such knowledge is still far from established. We have a number of candidate causes instead of necessary or sufficient causes, and worse, those candidate causes operate in a changing environment. The policy maker will be interested in how variables of interest react to changes in policy variables (instruments). Hence, the interest in causal inference is inspired by the desire to intervene. This notion is absent in the interpetations of causality given so far (although Cartwright, 1989, chapter 4, implicitly considers this issue).

Leamer (1985) discusses how causality should be related to intervention. He arrives at the familiar notion of structural parameters (where a parameter is said to be structural if it is invariant to a specified family of interventions). A causal relation is one with structural parameters, for a specific context of explicitly stated policy interventions. Leamer's definition adds an important element to Feigl's 'predictability according to a law'. Without identifying the relevant intervention, Feigl's definition remains sterile. But without experimentation, the notion of intervention is obscure as well. A possible help may be the identification of great historical events, such as wars and crashes. Leamer's view is that the best way to learn causes is from studying history, econometrics being of secondary interest only.

To summarize, while Wold imposes causality *a priori*, and Granger proposes to test for causality by means of a prediction criterion, Feigl's predictability according to a law takes a sensible position in between. Zellner (1979) recommends Feigl's approach to causality, but causality is only of interest if the law is relevant to a specific intervention. The probable effect of the intervention is of interest to the policy maker. This is a problem of probabilistic inference, not a metaphysical speculation about the ontological meaning of cause or the causal principle.

5 Science is prediction

5.1 Prediction as aim

5.1.1 Prediction and inference

Prediction may help to establish causality. But some investigators, in particular positivists, go further: they argue that prediction by itself is the ultimate aim of science.[17] Comte, for example, writes that prediction is the 'unfailing test which distinguishes real *science* from vain *erudition*' (cited in Turner, 1986, p. 12) – erudition, that is plain description. A similar statement can be found in the writings of Ramsey (1929, p. 151), who argues:

> As opposed to a purely *descriptive* theory of science, mine may be called a *forecasting* theory. To regard a law as a summary of certain facts seems to me inadequate; it is also an attitude of expectation for the future. The difference is clearest in regard to chances; the facts summarized do not preclude an equal chance for a coincidence which would be summarized by and, indeed, lead to a quite different theory.

Ramsey's theory of science aims at prediction, as does Carnap's theory of induction, and many other philosophies of science.

Econometricians tend to praise the predictive aim. Koopmans (1947, p. 166) argues that prediction is, or should be, the most important objective of the study of business cycles.[18] Why prediction is so important is not always made clear. Sometimes, prediction is viewed as a means for social engineering or control (Marschak, 1941, p. 448; Whittle, [1963] 1983). Others suggest that prediction is a prerequisite for testing, and science is about testing (Friedman, Blaug). Koopmans invokes prediction to demarcate superficial relations (those of Burns and Mitchell, 1946) from deep econometric relations (based on 'theory'). Koopmans (1949, p. 89) claims (but does not substantiate his claim) that those 'superficial relations' lack stability and are, therefore, bad instruments for predictions.

Another argument for prediction is that it underlies probabilistic inference. As Keynes and Fisher argue, statistics has a descriptive and an inductive purpose. If description were the sole aim, it would be vacuous as there would be no reason to 'reduce' the data. The latter becomes important if one wants to make inferences about new observations. Here, the modified simplicity postulate becomes relevant, as it is a way of dealing with the trade-off between accurate description and simplicity. This trade-off is based on an inductive theory of prediction. The theory of simplicity is based on Bayesian updating, a special version of prediction.

A radical point of view is expressed by Thomas Sargent. In the preface to Whittle ([1963] 1983, p. v), Sargent claims that 'to understand a phenomena is to be able to predict it and to influence it in predictable ways'. This reflects the symmetry thesis of the logical positivists Hempel and Oppenheim (see Blaug 1980, pp. 3–9), which states that prediction is logically equivalent to explanation. Whittle is more cautious than Sargent: prediction can be based upon recognition ('a compilation of possible histories' analogous to the event to be predicted) of a regularity as well as upon explanation of the regularity (Whittle, [1963] 1983, p. 2). Explanation does not entail prediction, and one can extrapolate without an underlying explanatory model. Prediction, Whittle argues, is rarely an end in itself. For example, a prediction of next year's GNP may have as its purpose the regulation of the economy. Here, the goal of inference is not to optimize the accuracy of the prediction, but to optimize (for example) a social welfare function.

5.1.2 Fishing for red herrings: prediction, novel facts and old evidence

In his *Treatise on Light* (published in 1690) (alternatively, cited in Giere, 1983, p. 274), Christiaan Huygens argues that the probability of a hypothesis increases if three conditions are satisfied. First, if the hypothesis is in agreement with observations, secondly, if this is the case in a great number of instances, and most importantly,

> when one conceives of and foresees new phenomena which must follow from the hypothesis one employs, and which are found to agree with our expectations.

This view remains popular today, for example in the hypothetico-deductive philosophy of science (Carnap, Hempel) and in the conjecturalist version of it (Popper, Lakatos).

Confirmations of predictions of novel facts increase the support for theories. For this reason, prediction is an important theme in discussions on the methodology of economics. Not only Friedman (1953) argues that prediction is essential for appraising theories. Theil (1971, p. 545), Zellner (1984, p. 30) and Blaug (1980) concur. There is a dissonant voice in this choir, though: John Maynard Keynes, who states his opinion in a discussion of the views of Peirce. In 1883, Peirce wrote an essay on 'a theory of probable inference', in which he states the rule of predesignation: 'a hypothesis can only be received upon the ground of its having been verified by successful prediction' (Peirce, 1955, p. 210). Peirce means to say that one has to specify a statistical hypothesis in advance of examining the data (whether these data are new or already exist is irrelevant, this seems to be why he prefers predesignation to prediction). Successful novel predictions are necessary and sufficient for confirming a theory. Keynes ([1921] CW VIII, p. 337) does not agree:

> The peculiar virtue of prediction or predesignation is altogether imaginary. The number of instances examined and the analogy between them are the essential points, and the question as to whether a particular hypothesis happens to be propounded before or after their examination is quite irrelevant.

He continues:

> [the view] that it is a positive advantage to approach statistical evidence without pre-conceptions based on general grounds, because the temptation to 'cook' the evidence will prove otherwise to be irresistible, – has no logical basis and need only be considered when the impartiality of an investigator is in doubt. (p. 338)

Keynes distances himself from Peirce on the one hand, and the 'truly Baconian' approach of Darwin on the other hand. The point relates to a problem in Bayesian inference, i.e. how to deal with old evidence. If evidence has been observed already, how can it be interpreted probabilistically? According to one interpretation such evidence has a probability value equal to one (it occurred). If so, old evidence cannot confirm a theory. This follows from Bayes' principle, if $P(e) = 1$:

$$P(h|e) = \frac{(P(h) \cdot P(h|e)}{P(e)} = P(h). \tag{7}$$

However, in applications of Bayesian inference, one does not set $P(e)$ equal to one. Instead, $P(e)$ figures as a normalizing constant by which the left hand side in (7) satisfies the axiom that probabilities sum to one.

In my view, the problem of old evidence is a philosophical red herring. The contributions in Earman (1983), devoted to this issue, show that its consumption is not nourishing. However, data-mining (fishing, it is sometimes called) is a problem for inference, whether Bayesian, frequentist or not probabilistic at all. In his discussion with Tinbergen, Keynes complains that he is not sure whether Tinbergen did not 'cook' his results, even though he had little reason to suspect the impartiality of Tinbergen. Tinbergen's methods were rejected as a valid base for testing business cycle theories. One reason is the lack of homogeneity. A second reason, more clearly formulated by Friedman (1940, p. 639), is that empirical regression equations are 'tautological reformulations of *selected* economic data' and as such not valid for testing hypotheses. The regression equations remain useful for deriving hypotheses. The real test is to compare the results with other sets of data, whether future, historical, or data referring to different regions. Like Keynes, most econometricians accept this view.

Friedman's point is (implicitly) supported in the survey on prediction by the statisticians Ehrenberg and Bound (1993) as well. They note that textbooks on statistics, and many papers in the literature on probability theory, abound with claims like 'least squares regression yields a "best" prediction', where 'best' is defined according to some statistical criteria (e.g. minimum mean square error). They complain that 'statistical texts and journals are obsessed with purely deductive techniques of statistical inference' (p. 188). Although not unimportant, this preoccupation with statistical criteria tends to blur the importance of specification uncertainty. What is needed is less focus on 'best' and more effort to try models in 'many sets of data' (MSOD) rather than single sets of data (SSOD). It is essential to vary the conditions in which models are applied. This yields more support to *ceteris paribus* conditions, hence more reliable inference. What is of interest is to widen the range of conditions under which a prediction stands up.

5.2 The Austrian critique on prediction and scientism

Not all phenomena can be predicted: many are unique or not reproducible. Economists in the Austrian tradition are characterized by a fundamental scepticism of the positivist aims of observation, measurement and prediction. These economists reject all that econometricians stand for. In his exposition of Austrian methodology, Israel Kirzner (1976, p. 43) argues that

> Our dissatisfaction with empirical work and our suspicion of measurement rest on the conviction that empirical observations of past human choices will not yield any regularities or any consistent pattern that may be safely extrapolated beyond the existing data at hand to yield scientific theorems of universal applicability.

This is as radical a scepticism as scepticism can be. It goes well beyond Humean scepticism (Hume argued that we merely observe regularities, but cannot infer causal relations from these). Although some economists claim universal applicability for their 'laws', most economists are more modest. Kirzner attacks a straw man, like Robbins did before.

Kirzner's argument is weak for another reason as well. The fact that human action is to some degree erratic does not imply that it is wholly unpredictable, either on the individual level or in the aggregate. If there is no scope for predictions, human action is impossible. One might invoke Hume's view that we cannot help but make inductions, or Keynes, who praises convention, or philosophers of probability, who argue that it is rational to make inferences and predictions. Although we are not absolutely certain that the sun will rise tomorrow, or that consumption of gas declines as its price rises, this does not prevent us from making predictions and acting upon those predictions. The Austrian doctrine is that we do have knowledge on the effect of price changes on gas consumption, but this knowledge results from introspection. This is the Kantian *a priori* response to scepticism.

Hayek (1989) is more subtle than Kirzner. Although Hayek criticizes the pretence of knowledge that economists tend to have, he argues that pattern prediction is feasible (and in 1933, the 'young' Hayek even argued that statistical research is 'meaningless except in so far as it leads to a forecast'; cited in Hutchison, 1981, p. 211). Pattern prediction should be distinguished from prediction of individual events, which is not possible. Hayek warns that we should not use pattern predictions for the purpose of intervention, as interventions are unlikely to yield the desired effects.[19] This is a hypothesis, though, not a fact, and econometric inference may serve to appraise the hypothesis.

Mathematical reasoning and algebraic equations may be helpful in gaining knowledge of the economy, but one should be careful in supplementing these tools with statistics. This leads to the illusion, Hayek (1989, p. 5) writes,

> that we can use this technique for the determination and prediction of the numerical values of those magnitudes; and this has led to a vain search for quantitative or numerical constants.

The pretence of knowledge lies in the claim that mathematical models, supplemented by numerical parameters (obtained with econometric tools) can be used to intervene in the economy and predict or obtain specific goals. Hayek is sceptical about the achievements of econometricians, as (he argues) they have made no significant contribution to the theoretical understanding of the economy. But one could also accuse Hayek, and, more generally, economists working in the Austrian tradition, of a very different pretence of knowledge. Their claim, that we may trust an 'unmistakable inner voice' (an expression due to Friedrich von Wieser, cited in Hutchison, 1981, p. 206), easily leads to dogmatism – is it not a pretence of knowledge, that our inner voice is unmistaken?

Caldwell, the advocate of methodological pluralism, accepts the Austrian critique of forecasting for an odd reason. Austrians reject prediction as a way of testing theories because of the absence of natural constants. If we want to evaluate Austrian economics, Caldwell (1982, p. 123) writes, then we should 'focus on the verbal chain of logic rather than on the predictions of the theory'. Hence, we should not engage in 'external criticism' (p. 248). By the same argument, we must accept that Friedman's monetarism should be evaluated on ground of its predictions. You cannot have it both ways: to evaluate monetarism on the basis of predictions, while evaluating Austrian economics on the basis of introspection or 'sound reasoning'. The modified simplicity postulate takes consideration of 'sound reasoning' as well as predictive performance. It does not depend on unjustified pluralism, but on a proper trade-off between the different ingredients of scientific arguments.

6 Testing

Testing belongs to the basic pastimes of econometricians. A casual investigation of titles of papers shows that there is a lot of 'testing' in the econometric literature, though not quite as much 'evidence'.[20] Why do econometricians test so much? And what is its significance? Sometimes one wonders about the abundance of tests reported in empirical papers as the purpose of all those tests is not always communicated to the reader. The search for homogeneity in consumer demand illustrates the frequent irrelevance of a positive or negative test result.

Statistical inference is usually understood as estimation and testing. In the following, I discuss some aims and methods of testing.

6.1 Why test?

6.1.1 Theory testing

The founder of modern statistics, R. A. Fisher ([1925] 1973, p. 2) wrote: 'Statistical methods are essential to social studies, and it is principally by the aid of such methods that these studies may be raised to the rank of sciences.' Fisher nowhere discusses the problem of testing rival theories, which would be the Popperian aim. There is, however, an episode in his research activities in which he was involved in a real scientific dispute: the question whether smoking causes lung cancer (see Fisher Box, 1978, pp. 472–6).[21] Confirmative evidence came from a large-scale study performed in Britain in 1952: it revealed a clear positive association between lung cancer and smoking. Fisher challenged the results: correlation does not prove causation. Fisher's alternative was that there might be a genetic cause of smoking as well as of cancer.

How can these rival theories be evaluated? Fisher's argument is one of omitted variables. One either needs information on the disputed variable in order to make correct inferences, or the experimental design should be adapted. In the latter case, one could think of constructing two randomized groups, force members of group I to smoke and force the others to refrain from smoking. This is not practical. Fisher, the geneticist, resorted to an alternative device: identical twins. As a research technique, it resembles the 'natural experiments' discussed in chapter 6. It did not resolve the dispute.

Even in this conflict, Fisher had positivist aims: measurement of parameters. The influence of Popperian thought resulted in a shift from estimating to testing as the hallmark of science. Prominent members of the Cowles Commission, like Haavelmo and Koopmans, advocated this approach. More recently, hypothetico-deductive econometrics (see Stigum, 1990) has been embraced by followers of the new-classical school, who search for 'deep' (structural) parameters. A characteristic view of a mainstream econometric theorist (Engle, 1984, p. 776), is that if 'the confrontation of economic theories with observable phenomena is the objective of empirical research, then hypothesis testing is the primary tool of analysis'.

The mainstream view is not without critics. Hahn (1992) writes, 'I know of no economic theory which all reasonable people would agree to have been falsified.' This is a theorist's challenge to empirical researchers. Hahn's verdict is that we need more thinking (theory), less empirical econometrics. McCloskey (1985, p. 182) agrees with Hahn's view: 'no proposition about economic behaviour has yet been overturned by econometrics'. Summers (1991, p. 133) adds, '[i]t is difficult to think today of many empirical studies from more than a decade ago whose bottom line was a parameter estimate or the acceptance or rejection of a hypothesis'. Summers argues that the formalistic approach to theory testing, in the tradition of Cowles and elaborated by new-classical econometricians, leads merely to a 'scientific illusion'.

Formal econometric hypothesis testing has an unpersuasive track record. One of the few econometricians who explicitly acknowledge this is Spanos (1986, p. 660):

> no economic theory was ever abandoned because it was rejected by some empirical econometric test, nor was a clear-cut decision between competing theories made in lieu of such a test.

But despite the lack of success, Spanos still aims at theory testing. He claims that it can be achieved by more careful specification of statistical models and relies strongly on mis-specification testing.

An important reason for the popularity of theory testing is that it is thought to be a major if not the main ingredient of scientific progress (Popper, [1935] 1968; Stigler, 1965, p. 12; Blaug, 1980), the best way to move from alchemy to science (Hendry, 1980). Falsificationism has had a strong impact on the minds of economists. Popper is about the only philosopher of science occasionally quoted in *Econometrica*. In the philosophy of science literature, however, falsificationism has become increasingly unpopular. Not in the least because actual science rarely follows the Popperian maxims. As Hacking (1983, p. 15) notes, 'accepting and rejecting is a rather minor part of science' (see also the contributions in Earman, 1983, and those in De Marchi, 1988).

Insofar as theory testing is an interesting aim at all, it is not yet said that econometrics is the best tool for this purpose. Identifying informative historical episodes or devising laboratory experiments (increasingly popular among game theorists, who rarely supplement their experiments with statistical analysis, as casual reading of such experimental reports in *Econometrica* suggests) may generate more effective tests than Uniformly Most Powerful tests. In the natural sciences, theory testing rarely results from sophisticated statistical considerations. Giere (1988, p. 190) discusses the different attitudes towards data appraisal in nuclear physics and the social sciences. Nuclear physicists seem to judge the fit between empirical and theoretical models primarily on qualitative arguments. Test statistics such as χ^2 are rarely reported in nuclear physics papers contained in, for example, the *Physical Review* (Baird, 1988, makes a similar observation). Hence, theory (or hypothesis) testing clearly does not always depend upon the tools we learned in our statistics course.

6.1.2 Validity testing

A different kind of testing, ignored in philosophical writings but probably more important in actual empirical research than theory testing, is validity testing (mis-specification or diagnostic testing). Validity tests are performed in order to find out whether the statistical assumptions made for an empirical model are credible. The paper of Hendry and Ericsson (1991) is an example of extensive validity testing. These authors argue that, in order to pursue a theory test, one first has to be sure of the validity of the statistical assumptions that are made. In their view, validity testing is a necessary pre-condition to theory testing. However, even if theory testing is not the ultimate aim, validity testing still may be important. Much empirical work aims to show that a particular model (formally or informally related to some theory) is able to represent the data. If much information in the data remains unexploited (for example, revealed by non-white-noise residuals), this representation will be suspect or unconvincing to a large part of the audience.

The importance of validity tests should not be over-emphasized, at least, they should be interpreted with care. One may obtain a very neat 'valid' statistical equation of some economic phenomenon, after extensive torturing of the data. Such a specification may suggest much more precise knowledge than the data actually contain. Sensitivity analysis, for example along the lines of Leamer (1978) (see chapter 7), or such as performed by Friedman and Schwartz (1963), is at least as important as validity testing in order to make credible inferences. Illuminating in this context is the exchange between Hendry and Ericsson (1991) and Friedman and Schwartz (1991), discussed in detail in Keuzenkamp and McAleer (1999).

6.1.3 Simplification testing

A third aim of testing is simplification testing. Simple models that do not perform notably less than more complex ones are typically preferred to the complex ones (see chapter 5). Inference conditional on exogeneity assumptions is often preferred to analysing 'romantic' (Bunge, [1959] 1979) models where everything depends on everything and for which Walrasian econometricians advocate full information methods (Haavelmo and Koopmans belong to this category). Simplicity not only serves convenience and communication, but is needed for an epistemological theory of inference.

6.1.4 Decision making

Finally, a frequently expressed goal of testing is decision making. This view on testing, and its implementation in statistics, is primarily due to the Neyman–Pearson (1928; 1933a, b) theory of inductive behaviour. The decision-theoretic approach of testing for the purpose of inductive behaviour has been further elaborated by Wald and, from a Bayesian perspective, by Savage ([1954] 1972). Lehmann ([1959] 1986) is the authoritative reference for the frequentist approach, while Berger ([1980] 1985) provides the Bayesian arguments.

Decision making, based on statistical acceptance rules, can be important for process quality control, but may even be extended to the appraisal of theories. This brings us back to theory testing as a principal aim of testing. Lakatos claims that the Neyman–Pearson version of theory testing 'rests completely on methodological falsificationism' (Lakatos, 1978, p. 25 n.). In chapter 3, section 4.4, I criticized this view as historically incorrect and analytically dubious.

6.2 Methods of testing

6.2.1 General remarks

The most familiar methods of testing in econometrics are based on the work of Fisher, and on the Neyman–Pearson theory of testing. To summarize the relevant issues that were discussed in chapter 3, Fisher's theory of significance testing is based on the following features. First, it relies on tail areas (P-values). Secondly, it is intended for small samples. Thirdly, it is meant for inductive scientific inference. These features are clearly different from those of the Neyman–Pearson theory of testing. This is based, first, on an emphasis on size and power, leading to UMP tests. Secondly, Neyman–Pearson tests are meant for the context of repeated sampling. Finally, they are instruments for inductive behaviour and making decisions.

In chapter 3, I analysed some shortcomings of both methods. Despite these shortcomings, there are few close rivals to the blend of Fisher's and Neyman–Pearson's methods that pervaded econometrics. Engle (1984), for example, gives an overview of test procedures, all based on Neyman–Pearson principles. Bayesian testing has become popular only in a few specific cases (e.g. the analysis of unit roots). Importance testing (practised by Tinbergen and, in different forms, advocated by Leamer and Varian) is not very popular. The implementation of Neyman–Pearson methods at the practical level is not easy, though. There is a wide divergence between empirical econometrics and the maxims of a 'celibate priesthood of statistical theorists', as Leamer (1978, p. vi) observes. The popularity of data-mining is particularly hard to combine with the equally popular Neyman–Pearson based methods advocated in textbooks and many journal papers. Hendry (1992, p. 369) rightly notes that test statistics can be made insignificant by construction, as residuals are derived, not autonomous processes. The problem of interpreting the resulting test statistics remains unsolved today (see Godfrey, 1989, p. 3; also Leamer, 1978, p. 5).

The following sections deal more explicitly with alternative statistical methods to test rival theories: both from a frequentist and from a Bayesian perspective.

6.2.2 Frequentist approaches to testing rival theories

In the frequentist approach there are three different strategies for testing rival models: the embedding or comprehensive model strategy, the generalized likelihood ratio strategy and the symmetric (or equivalence) strategy. These approaches will be discussed below.

Comprehensive testing The comprehensive model is a nesting of the competing models in a more general, embedding model. Two rival models M_1 and M_2 have been formulated with reference to an $N \times 1$ regressand vector y. Assume both are linear models, represented by:

$$M_i\colon y = W\gamma_i + X_i\beta_i + u_i, \;\; u_i \sim N(0, \sigma_i^2 I), \;\; i = 1, 2, \qquad (8)$$

where $W(N \times K_w)$ is a regressor matrix common to both models ($K_w \geq 0$), and X_i are $N \times K_i$ regressor matrices unique to either model. Testing these hypotheses can be done by regarding them as two restricted forms of a more general model, the embedding model:

$$M^*: y = W\gamma^* + X_1\beta_1^* + X_2\beta_2^* + u^*, \;\; u^* \sim N(0, \sigma_i^2 I). \tag{9}$$

The simplest test is to test the hypotheses H_i: $\beta_i^* = 0$ (using an F-test, for example; see MacKinnon, 1983, pp. 95–6 for discussion and comparison with other tests). A difficulty is that there are four possible test outcomes: the combinations of rejections or acceptances of the hypothesis that $\beta_i^* = 0$. If either H_1 or H_2 is rejected, and the other accepted, then we have a 'desirable' test outcome that provides the required information. If both hypotheses are rejected (or accepted), the researcher enters a state of confusion. This may become even worse, if, in addition, a joint hypothesis $\beta_1^* = \beta_2^* = 0$, is tested, as the outcome of this hypothesis test may be in conflict with the separate tests (see Gaver and Geisel, 1974, p. 56).

A further complication of the comprehensive testing approach arises from the fact that the embedding model presented here is not the only conceivable one. One can also construct a comprehensive model by means of a weighted average of the two models. In fact, there are numerous ways to combine rival models. For example, by considering the likelihood functions of M_1 and M_2, one can obtain the comprehensive likelihood function $L(L_1, L_2, \lambda)$ as an exponential mixture (due to A. C. Atkinson but already suggested by Cox, 1961, p. 110), or as a convex combination (due to Quandt), etc. The exponential mixture is particularly convenient, as (in the univariate linear model) it yields the comprehensive model (with W for simplicity incorporated in the matrices X_i),

$$y = (1 - \lambda)X_1\beta_1 + \lambda X_2\beta_2 + u. \tag{10}$$

One can test whether λ differs from zero or one, but a problem is that this parameter is not always identifiable. If λ is exactly equal to one of those extremes, the parameters of one of the sub-models do not appear in the comprehensive likelihood function which invalidates the testing procedure (see Pesaran, 1982, p. 268).

A final problem of comprehensive testing is that the test properties (in terms of Type I and Type II errors) are unknown if the variables are non-stationary (and, of course, more generally if the models result from a specification search).

The generalized likelihood ratio test An alternative to the comprehensive model approach is due to Cox (1961, 1962), who provides an analysis of tests of 'separate' families of hypotheses based on a generalization of the Neyman–Pearson likelihood ratio test. To define terms, consider again the models M_1 and M_2. By separate families of hypotheses, Cox (1962, p. 406) refers to models that cannot be obtained from each other by a suitable limiting approximation. Vuong (1989) gives more precise definitions. He calls this the *strictly non-nested* case, defined by the condition,

$$M_1 \cap M_2 = \oslash. \tag{11}$$

As an example, one may think of two regression models with different assumptions regarding the stochastic process governing the errors (e.g. normal and logistic). The *comprehensive* approach, on the other hand, relies on nested models, defined by

$$M_1 \subset M_2. \tag{12}$$

An intermediate case, which is the most interesting for econometrics, consists of *overlapping* models, with

$$M_1 \cap M_2 \neq \oslash, \;\; M_1 \not\subset M_2, \;\; M_2 \not\subset M_1. \tag{13}$$

I will concentrate on the case of overlapping models. First note that the Neyman–Pearson likelihood ratio test does not apply in this case. If L_i denotes the maximized likelihood of M_i, one can calculate the likelihood ratio,

$$L_{i/2} = \frac{L_i}{L_i}. \tag{14}$$

The likelihood ratio test,

$$LR = -2\ln\left(\frac{L_1}{L_2}\right) \tag{15}$$

has an asymptotic χ^2 distribution under the null, model M_1 for example. However, this is only the case if $M_1 \subset M_1$ (as $\chi^2 \geq 0$). In other words, the plain likelihood ratio test can be applied in the context of the comprehensive approach, but not in cases where one wants to evaluate 'separate' families of hypotheses or overlapping models. If one is not particularly interested in significance testing but in model discrimination, then it may still be useful to consider the maximized likelihoods and select the model with the higher one (this is also noted in Cox, 1962), perhaps after considering matters such as simplicity using the modified simplicity postulate. But Cox aims at constructing a proper significance test in the spirit of the Neyman–Pearson framework in the absence of prior probability distributions.[22] This test is an elaboration of the simple likelihood ratio test (Cox, 1961, p. 114):

$$LRC = LR_{1/2} - \mathcal{E}_1(L_{1/2}), \tag{16}$$

where $L_{1/2}$ is the log-likelihood ratio, $\mathcal{E}_1\{L_{1/2}\}$ is the expected value (if M_1 were true) of the difference between the sample log-likelihood ratio given M_1 and the sample log-likelihood ratio given M_2 (the likelihoods are evaluated at the maximum likelihood estimates). Cox (1961) shows that under the null, LRC is asymptotically normal.

An essential feature of this procedure is that it is asymmetric: inference may depend on whether M_1 or M_2 is chosen as the reference hypothesis, and it may happen that M_1 is rejected if M_2 is the reference model and vice versa, in which case there is inconsistent evidence. If the problem is not one of model selection but specification search, this may be problematic: it gives a hint that 'something' is wrong.

The Cox test first appeared in econometrics in the early 1970s (see, e.g., the surveys by Gaver and Geisel, 1974, and MacKinnon, 1983). Hendry, Mizon and Richard use it in their theory of encompassing, which is used for mis-specification analysis as well as testing rival theories (see, e.g., Mizon and Richard, 1986). In practice, they rely on linear embedding models with simple tests for adding a subset (excluding the overlapping variables) of variables of a rival model to the reference model. For this purpose, a set of t-tests or an F-test (see the discussion of embedding, above) can be used ('parameter encompassing'), plus a one degree of freedom test on the variances of the rival models ('variance encompassing').

Symmetric testing A different frequentist approach to the problem of model choice is presented in Vuong (1989), who drops the assumption that either one or both models should be 'true' in the statistical sense. This is the third strategy in the frequentist domain, the symmetric or equivalence approach (which dates back to work of Harold Hotelling in 1940). Vuong provides a general framework in which nested, non-nested or overlapping models can be compared, with the possibility that one or both of the models being compared may be mis-specified.

He uses the Kullback Leibler Information Criterion (KLIC), which measures the distance between a given distribution and the 'true' distribution. The idea is 'to define the "best" model among a collection of competing models to be the model that is closest to the true distribution' (Vuong 1989, p. 309).

Let the conditional density functions of the 'true' model be given as $h^0(y_t|z_t)$. The data $x_t = (y_t, z_t)$ are assumed to be independent and identically distributed, therefore, the approach is not directly applicable to time series models. The 'true' model is not known, but there are two alternative models, parameterized by θ and γ respectively. If they do not coincide with the 'true' model, they are mis-specified. In this case, one can use quasi-maximum likelihood techniques to obtain the 'pseudo-true' values of the parameters, denoted by an asterisk (see Greene, 1993, for discussion of this method). Given the vectors of observations $\{y, z\}$, the KLIC for comparing the distance between the 'true' model and one of the alternatives, e.g. the one parameterized by θ, is defined as:

$$KLIC(H^0_{y|z}; M_\theta) = \mathcal{E}^0\left[\ln f^0(y|z)\right] - \mathcal{E}^0[\ln f(y|z; \theta^*)], \tag{17}$$

where M_θ denotes the model with parameterization θ and $f^0$is the 'true' conditional density. As the true model is unknown, this measure cannot be calculated. However, if the distance between the 'true' model and the alternative is defined similarly, we can define an indicator I as:

$$I = \mathcal{E}^0[\ln f(y|z); \theta^*)] - \mathcal{E}^0[\ln f(y|z; \gamma^*)]. \tag{18}$$

Again, it is unknown if the 'true' model is not known. However, under certain regularity conditions and given the existence of unique 'pseudo-true values' for θ and γ, it can be estimated consistently using the likelihood ratio statistic (Vuong, 1989, lemma 3.1). If the models do not coincide, the test statistic (the value of this indicator) goes to plus or minus infinity almost surely, depending on which of the two models is closest to the 'true' model. In case of equivalence of the models (i.e. $f(y|z; \theta^*) = f(y|z; \gamma^*)$, the test has a weighted sum of χ^2 distributions.

Vuong's procedure has clear advantages: it does not rely on the axiom of correct specification. The advantage of generality is countered by a loss of power to discriminate between the models, which, together with the computational complications, is a drawback of this approach. Also, the restriction to the independent and identically distributed case is a drawback. Finally, the reliance on quasi-maximum likelihood relies strongly on asymptotic reasoning. In many economic applications, one has to deal with particularly small samples. Exact small sample statistics are, therefore, more desirable than tests based upon sophisticated asymptotic theory of which the small sample properties are not known.

6.2.3 Bayesian approaches to testing rival theories

The second methodology is based on the Bayesian perspective. There are at least two approaches within the Bayesian methodology: one based on predictive probability density functions, the other on decision theory.

The predictive approach The predictive approach to inference is based on the idea that inference should be based on observables only. Data are observable, parameters of models are not. Hence, appraisal of rival models should be based on the data, not on unknowns. The basic idea is very simple. Instead of applying Bayes' theorem to infer about parameters, we now apply it to inference of models. More explicitly, assume, as before, we have two models to compare: M_1 and M_2, parameterized by β_1 and β_2 respectively. These models are designed to describe data y. The prior probability density functions for the parameters are $f(\beta_i|M_i)$, $i = 1, 2$. The predictive probability density functions of the observations given the models, $f(y|M_i)$, is obtained by:

$$f(y|M_i) = \int f(y|\beta_i, M_i) f(\beta_i|M_i) d\beta_i, \tag{19}$$

where $f(y|\beta_i, M_i)$ is the familiar likelihood function. To obtain information on the comparative credibility of the rival models given by the data, one can calculate $f(M_i|y)$. This is a straightforward application of Bayes' theorem. The additional ingredients are the prior probabilities of the models, denoted by $f(M_i)$. If, *a priori*, both models are equally credible, it is natural to set $f(M_i)$ equal to 0.5. These probabilities are revised in the light of the data to posterior probabilities:

$$f(M_i|y) = \frac{f(M_i) f(y|M_i)}{f(y)}. \tag{20}$$

The predictive density of future observations, y_f, given oberved data y, can be obtained by the predictive densities of those observations as given by either of the rival models, weighed by the credibility of each model:

$$f(y_f|y) = f(M_1|y) f(y_f|y, M_1) + f(M_2|y) f(y_f|y, M_2), \tag{21}$$

where $f(y_f|y, M_i)$ is obtained by integrating out the parameters of model M_i from the predictive density of new observations (weighted by the prior probability density function (pdf) of the parameters).

The predictive approach to inference does not force one to *choose* between the models. Optimal inference on future data depends on both (two, or even more) models. Evidence is combined, not sieved. This is a strength as well as a weakness. It does not make sense to invest in calculating the predictive pdf obtained by models that are quite off the mark (such a model would carry a low weight in the combined prediction). Similarly, parsimony suggests that if both models yield similar predictive results, one might just as well take only one of the models for making a prediction. Hence, it may be desirable to make a choice. This can be done by extending the predictive approach with a decision theoretic-analysis.

The decision theoretic approach The decision theoretic approach to the evaluation of rival models is based on the posterior odds approach, due to Jeffreys ([1939] 1961). The marginal densities of the observations for the rival models are compared by means of the posterior odds ratio,

$$POR = \frac{f(M_1|y)}{f(M_2|y)} = \frac{f(M_1)}{f(M_2)} \cdot \frac{f(y|M_1)}{f(y|M_2)}. \qquad (22)$$

The posterior odds ratio is equal to the prior odds times the ratio of the weighted or averaged likelihoods of the two models. This is the so-called Bayes factor, the ratio of the marginal densities of y in the light of M_1 (with parameters β_1) and M_2 (with parameters β_2) respectively. The Bayes factor reveals the quality of the representations.

The averaging of the likelihoods is the result of the uncertainty about the unknown parameters β_i. The posterior odds ratio is obtained by integrating this uncertainty out:

$$POR = \frac{f(M_1|y)}{f(M_2|y)} = \frac{f(M_1)}{f(M_2)} \cdot \frac{\int f(y|\beta_1, M_1) f(\beta_1|M_1) d\beta_1}{\int f(y|\beta_2, M_2) f(\beta_2|M_2) d\beta_2}. \qquad (23)$$

This expression makes clear how Jeffreys' posterior odds approach relates to the Neyman–Pearson likelihood ratio test. The latter is based exclusively on the fraction $f(y|\beta_1, M_1)/f(y|\beta_2, M_2)$, evaluated at the maximum likelihood estimates – hence, it only depends on the goodness of fit of the models. There is no consideration of the prior odds of the models, unlike in the posterior odds analysis. In case of comparing two linear regression models, Zellner (1971, pp. 310–11) shows that the posterior odds, on the other hand, depend on the prior odds of the rival models, precision of the prior and posterior distributions of the parameters, goodness of fit, and the extent to which prior information agrees with information derived from the data. The goodness-of-fit factor dominates if the sample size grows to infinity.

In practical applications, the Bayesian researcher usually sets this prior odds ratio equal to one, in order to give both models an equal chance. Private individuals can adjust this ratio with their subjective odds as they like. A more fundamental difference between the Neyman–Pearson approach and posterior odds is that the first is based on the maxima of the likelihood functions, while the latter is based on averages of likelihood functions. For averaging the likelihood functions, one needs prior probability distributions (or densities), a reason why some investigators do not like the posterior odds approach. One may interpret the difference as a dilemma in constructing index numbers (Leamer, 1978, pp. 100–8): few statisticians propose the highest observed price as the ideal 'index' for consumption goods' prices. The likelihood ratio test is like this 'index'. If likelihood functions are relatively flat (compare: most elements of the class to be summarized in the index have values close to the maximum), this may be not too bad an approximation to the ideal index. If, however, data are informative (for example, due to a large sample size), the Neyman–Pearson likelihood ratio test statistic and its relatives, Cox's test and Vuong's test, are unsatisfactory: the maximum values of the two likelihood functions are not representative for the likelihood functions as a whole. This leads to conflicting results between likelihood ratio tests and posterior odds ratios, if one hypothesis is a sharp null and the alternative a composite hypothesis.

Another distinction between the Neyman–Pearson approach and Jeffreys' posterior odds ratio is that the latter does not depend on hypothetical events but on the observations and prior probability densities.

The decision theoretic approach builds on Jeffreys' posterior odds by considering the possible losses of making wrong decisions (see Savage, [1954] 1972; Gaver and Geisel, 1974; Berger, [1980] 1985). An example is the MELO, or Minimum Expected Loss, approach. Consider the following very simple loss structure. Given the 'correctness' of either model M_1 or M_2, a loss structure with respect to the rival models might be that accepting a 'correct' model involves zero loss, while accepting the wrong model leads to a positive loss of L. A more complicated loss structure might be one that is based on prediction errors (see Zellner, 1984, p. 239, for discussion of squared error loss). The posterior expected loss of choosing model i and rejecting j for the simple example given here is

$$EL(M_i) = 0 \cdot f(M_i|y) + L \cdot f(M_j|y). \tag{26}$$

In general, the model with the lowest expected loss will be chosen. The Bayesian decision theoretic approach uses the posterior odds ratio but adds an appropriate loss structure to the decision problem. M_1 will be chosen if the weighted (averaged) likelihood ratio, $f(y|M_1)/f(y|M_2)$, exceeds the prior expected loss ratio.

7 Summary

The beginning of this chapter sketched the positivists' view on science. This view, I argued, is still relevant to econometrics. It is characterized by an emphasis on observation, measurement and verification. Furthermore, positivists do not believe in the need to search for 'true causes' and tend to be sceptical about deep parameters. They are Humeans, without embracing Hume's response to scepticism.

The aims of measurement, causal inference, prediction and testing seem to compete. If, however, we take a step back and reconsider the theory of simplicity, then it becomes possible to make order out of the chaos of rival aims. Description is useful as it serves prediction. Explanation is useful as well (here, some hard-line positivists might disagree), as it allows the use of analogy and the construction of a coherent theory (minimizing the number of auxiliary hypotheses), which is economic and again, this serves prediction. The final section on testing rival theories showed that the Bayesian approach allows for a very useful combination of prediction and testing. A great advantage is that it does not rely on the existence of one 'true' model.

The discussion on the meaning of causal inference for econometrics may be clarified by accepting Hume's scepticism, without concluding that 'everything would have been produced by everything and at random' (as Sextus Empiricus wrote). To say that, for example, the minimum wage causes unemployment, is 'convenient shorthand' for the rational expectation that an increase of the minimum wage will be accompanied by an increase in unemployment, and more importantly, that an intervention in the minimum wage with a 1% change will result in an x% change in unemployment. This is perfectly legitimate, even for a Humean sceptic. The rational expectation is supported by empirical research: measurement, inference, identification of 'historical experiments', perhaps even proper econometric tests. Conditional prediction is the principal aim of economics: in the end, it is the rational expectation of the effect of interventions that makes economic analysis relevant.

The aims of econometric inference can be summarized as: finding regularities that are simple as well as descriptively accurate, and that are stable with reference to intended interventions. Sceptics, who hold that econometrics has not yielded much interesting empirical knowledge, may be right when they search for 'realist' knowledge: truth. Econometrics does not deliver truth.

Notes

1. Pearson ([1892] 1911, p. 86) argues that the same is true in physics, e.g. Newton's law of gravitation.
2. Note that Friedman had 'long discussions' with Popper in 1947 at the founding meeting of the Mont Pelerin Society. Friedman found Popper's ideas 'highly compatible' with his own. In a pragmatic sense, Friedman is right: both advocate confrontation of predictions with the data. From a philosophical perspective, there are great differences, where Friedman is too modest when he writes that Popper's views are 'far more sophisticated' (see Friedman and Friedman, 1998, p. 215).
3. It is ironic that its prominent member, Koopmans, attacked the economists at the National Bureau of Economic Research who took this slogan most seriously: Burns and Mitchell. The emphasis of the Cowles Commission, from at least 1940 onwards, was on theory (the first input in the hypothetico-deductive model of science), not measurement.
4. Quetelet measured anything he could. An example is the blooming of lilacs in Brussels. He went beyond description, in this case. A law of lilacs is the result. Let t be the mean daily temperature (degrees Celsius). The lilacs will blossom if $\Sigma t^2 > 4264$°C, with t summed from the last day with frost onwards (see, Hacking, 1990, p. 62).
5. The problem with his realism on preferences can be exposed by asking whose kind of preferences are not doubted: Hahn's? Stigler and Becker's? Veblen's?
6. Caldwell (1982) notes that Robbins does not use the words *a priori* in his essay. Robbins' emphasis on the self-evident nature of economic propositions justifies my classification of Robbins' response to scepticism as the *a priori* strategy.
7. Lucas (1987, p. 45) does not claim that models should be 'true', they are just 'workable approximations' that should be helpful in answering 'a limited set of questions'.
8. It is difficult to define Kant's position. I will follow Russell (1946) who holds that Kant belongs to the rationalist tradition.
9. Note that Feigl considered causality both in the context where intervention is relevant and where it is not (astronomy). His definition encompassed both cases. Zellner at least implicitly has interventions in mind.
10. Another, more modern, approach is due to Mackie (1965), who analyses causation by means of INUS conditions (Suppes, 1970, pp. 75–6, extends

this to the probabilistic context). An INUS condition is a condition that is an Insufficient but Necessary part of a condition which is itself Unnecessary but Sufficient for the result. An example is: if a short circuit (A) is an INUS condition for fire (B), an additional condition C might be the presence of oxygen, and the event D would be another, but distinct, way by which the fire might have been caused. C and D serve as background information. An INUS condition does not have to be a genuine cause (see the discussion in Cartwright, 1989, pp. 25–7).

11. A similar statement of Kant's is that everything that happens 'presupposes something upon which it follows in accordance to a rule' (translation in Krüger, 1987, p. 72).
12. More precisely, if C happens, then (and only then) E is always produced by it.
13. Although Leibniz rejected gravitation as an 'inexplicable occult power', not a true cause (see Hacking, 1983, p. 46).
14. J. S. Mill's fifth canon of induction, the method of concomitant variation, says that if two variables move together, either is the cause or the effect of the other, or both are due to some (other) fact of causation.
15. In Cartwright's book, the meaning of 'capacity' gradually changes to autonomous relations, like the ones that generate the 'deep parameters' that are the principal aim of new-classical econometricians. I think stability and autonomy are different issues. I would agree with the statement that econometric inference for purposes of prediction and intervention needs relative stability, but I am not persuaded that the notion of capacity is needed as well.
16. The original definition of Granger (1969) is given in terms of variances of predictive error series, where roughly speaking $\mathbf{x}$ causes $\mathbf{y}$ if $\mathbf{x}$ helps to reduce this variance in a linear regression equation.
17. Occasionally it is acknowledged that not all great works in empirical science are of a predictive nature. Darwin's theory of evolution is a notorious example. Indeed, it is hard to predict future evolution of man or other animals. But Darwin's theory is not void of predictions: they are the so-called 'missing links', and discoveries such as the Java and Peking men are examples of confirmed predictions of Darwin's theory (van Fraassen, 1980, p. 75).
18. Note that in his earlier writings, he warns that prediction is a dangerous undertaking, as it may lead to speculation and instability (Koopmans, 1941, p. 180).
19. The Austrian aim of economics is to explain how unintended consequences result from purposeful actions. Intervention (government policy to stabilize the economy) will yield additional unintended consequences and should, therefore, be avoided.
20. A search with 'online contents' of all (not only economics) periodicals at the Tilburg University Library from January 1991 to 12 August 1993 (containing 212,856 documents) yielded the following results:
    2,702 papers have a title with 'testing' (and variations thereof) as an entry
    2,133 papers have a title with 'estimating' (and variations) as an entry
    92 papers have a title with 'rejection' (and variations) as an entry

62 have a title with 'confirmation' as an entry
1,443 papers have a title containing the word 'evidence'.

21. A similar dispute raged in 1911 and 1912 between Karl Pearson and Keynes about the influence of parental alcoholism on children. Pearson established that there was no effect, Keynes disputed Pearson's inference. According to Keynes, Pearson's analysis does not control for all the relevant factors: there was what might be called spurious non-correlation. Keynes' main objection to Pearson was ethical: it is very unlikely that alcoholism is irrelevant, you therefore should control for everything that possibly might be relevant before making statistical claims. It is immoral to make a wrong judgement in a case like this. See Skidelski (1983, pp. 223–7) for discussion.
22. If proper prior probability distributions are available, then the Bayesian posterior odds approach suggests itself. Cox (1961, p. 109) notes that if such distributions are not available, and use is made of improper distributions, then the odds are driven by the normalizing constants of M_1 and M_2. See the discussion of the Bayesian approach below for some comments.

10 Probability, econometrics and truth

> Your problems would be greatly simplified if, instead of saying that you want to know the 'Truth,' you were simply to say that you want to attain a state of belief unassailable by doubt.
>
> C. S. Peirce (1958, p. 189)

1 Introduction

The pieces of this study can now be forged together. The first chapter introduced the problems of induction and Humean scepticism. Different responses to scepticism were mentioned: naturalism, apriorism, conjecturalism and probabilism. In the remaining chapters, I analysed the foundations of a probabilistic approach to inference. Rival interpretations of probability were presented and their merits compared. I provided arguments as to why I think apriorism and conjecturalism are not satisfactory, although I do not deny that the probabilistic approach is problematic as well.

Chapter 3 discussed the frequency approach to probabilistic inference, chapter 4 considered the epistemological (Bayesian) perspective. Although the logical foundations of the frequency approach to probabilistic inference are problematic, probability theorists have been able to develop an impressive box of tools that are based on this interpretation. Unsatisfactory philosophical foundations do not necessarily prevent a successful development of a theory. On the other hand, the logically more compelling epistemological approach has not been able to gain a strong foothold in applied econometric inference. Econometricians have difficulties in specifying prior probability distributions and prefer to ignore them. Still, they tend to give an epistemological interpretation to their inferences. They are Bayesian without a prior.

The applications of frequentist methods in econometric investigations reveal, however, that there is a discomforting gap between the (frequentist) theory and practice of econometrics. This gap has lead to increasing scepticism about the accomplishments of econometric inference. Econometric practice is not the success story that the founders of econometrics expected. In 1944, Marschak, director of the Cowles Commission, wrote a memorandum to the Social Science Research

Committee (quoted in Epstein, 1987, p. 62), that stated his research agenda for the near future:

1945–6: work on method(ology) to be completed in the main
1946–8: final application of method to business cycle hypotheses and to (detailed) single market problems
1948–9: discussion of policy. Extension to international economics

A few years before, a similar agenda had worked for the development of the atomic bomb, but statistical evaluation of economic theories turned out not to be that easy, not to speak of social engineering.[1]

An old problem, specification uncertainty, subverted the credibility of econometric investigations. The econometrician's problem is not just to estimate a given model, but also to choose the model to start with. Koopmans (1941, p. 179) and Haavelmo (1944) delegate this problem to the economic theorists, and invoke the 'axiom of correct specification' (Leamer, 1978) to justify further probabilistic inference. This problem of specification uncertainty was noticed by Keynes (remember his call for an econometric analogy to the Septuagint). Econometricians have been unable to respond convincingly to Keynes' challenge. This has discredited the probabilistic response to Humean scepticism, and, perhaps for that reason, methodological writings in economics are not characterized by an overwhelming interest in econometric inference. Instead, falsificationism, apriorism and 'post-modernism' dominate the scene.

Section 2 of this chapter deals with the methodology of economics. I reconsider why the probabilistic response to scepticism deserves support. Which probability interpretation should underlie this probabilistic response, is the theme of section 3. In section 4, I discuss the temptations of truth and objectivity. Section 5 provides a summary.

2 Methodology and econometrics

2.1 Popperians without falsifications

Philosophers of science are not regularly quoted in *Econometrica*. If, however, an econometrician refers to the philosophy of science, it tends to be in praise of Popper or Lakatos. However, econometricians rarely try to falsify. Exceptions prove this rule. One exception is Brown and Rosenthal (1990, p. 1080). Without referring to Popper these authors, who test the minimax hypothesis, write that 'the empirical value of the minimax hypothesis lies precisely in its ability to be rejected by the data'. And indeed, they conclude that the hypothesis is rejected by the data. This is the Popperian way of doing science (albeit in a probabilistic setting) and, occasionally, it yields interesting results (in particular, if a rejection points where to go next, which is the case in the example cited). If one searches hard enough, one should be able to find additional examples of Popperian econometrics. However, they are rare. A similar conclusion was drawn in chapter 8 on the search for homogeneity.

Econometricians usually are neither Popperian nor Neyman–Pearson decision makers. They tend to search for adequate empirical representations of particular data. This is what positivism is all about (see chapter 9, section 2). Accepting an economic theory on the basis of econometric arguments 'involves as a belief only that it is empirically adequate' (van Fraassen, 1980, p. 12). However, empirical adequacy is too vague to serve as a guiding principle in econometrics. For example, if 'adequate' means 'fits well', one has to confront the problem that some empirical models cannot be less adequate than rival ones which are a special cases of empirical models, as Lucas pointed out when he compared (Keynesian) disequilibrium models to new-classical equilibrium models. Adequacy should be supplemented by parsimony and internal consistency ('non-*ad-hoc*ness'). The trade off can be investigated by means of the modified simplicity postulate.

2.2 Econometrics in the methodological literature

Econometricians, who are interested in the philosophy of science, would benefit more from reading the methodological works of Karl Pearson, Ronald A. Fisher and Harold Jeffreys, than from reading Popper, Lakatos or methodological writings on economics such as Blaug (1980) or Caldwell (1982). I will turn to the literature on the methodology of economics to show how it deals with econometrics. Few writings in the methodology of econometrics take a probabilistic perspective. I will briefly illustrate the fate of econometrics in methodological writings, by passing through the works of Blaug, Caldwell, Hamminga, Boland and McCloskey.

Blaug is a Popperian. He is aware that economists do not obey Popper's rules, but instead of concluding that Popper is wrong, he complains that econometricians are essentially lazy or negligent in not confronting theories with tough tests. Applied econometrics too often resembles 'playing tennis with the net down' (Blaug, 1980, p. 256). Blaug does not discuss the scope and limits of statistical inference. Apart from a few casual remarks about Neyman–Pearson (pp. 20–3), there is no analysis of the (practical or philosophical) difficulties that may arise in econometric testing. In this sense, Hendry's writings (that are claimed to follow the conjecturalist model of inference) are complementary to Blaug, and despite my disagreements with his view, he provides a better understanding of how econometricians try to test than Blaug does.

A second popular methodologist is Caldwell. Caldwell (1982, p. 216) complains that econometricians neglect, or only make gratuitous references to, philosophical issues. This is the only occasion in his book where the word econometrics occurs – the index of *Beyond Positivism* has no entry on related topics such as statistics, probability or Neyman–Pearson. Spanos (1986, chapter 26), tries to meet Caldwell's challenge to integrate economic methodology and econometrics. Spanos does not discuss Caldwell's views on apriorism, which has become a more dominant theme in later works of Caldwell.

Hamminga (1983, pp. 97–101), in a case study of the theory of international trade, discusses problems of identification and specification uncertainty in econometrics and concludes that they result inescapably in *ad hoc* procedures. He concludes that 'results of econometric research cannot in the least affect the dynamics of the Ohlin–Samuelson programme' (p. 100). This means that these two elements are independent of each other: the theoretical developments in the theory of international trade could have gone the same way if no econometric research had ever been done at all. If this claim holds universally, then we may wonder what the use of econometrics is. But I do not think Hamminga's claim has such universal status. The independence of these two disciplines suggests that econometric research could similarly continue without taking notice of theoretical developments, which, I think, is an entirely unwarranted conclusion. For example, Leamer (1984) provides an extensive discussion of the empirical literature as well as an econometric analysis of the Heckscher–Ohlin theory. Leamer does not ask whether this theory should be accepted or rejected, but rather how adequately it is able to represent the data. Another example is the theory of consumer behaviour, which was stimulated by the empirical findings (think of flexible functional forms, dynamic models, aggregation). A third example is the surge in theoretical analysis of the Permanent Income Hypothesis, largely driven by empirical econometrics (see the investigations in liquidity constraints and excess sensitivity).

A discussion of econometrics similar in scope and contents to Caldwell's can be found in the well-known book of Boland (1982, p. 4). He briefly turns to econometrics, and concludes:

> Presentations of methodology in typical econometrics articles are really nothing more than reports about the mechanical procedures used, without any hint of the more philosophical questions one would expect to find in a book on methodology.

I prefer to turn this upside down. Presentations of econometrics in typical methodological articles are really nothing more than badly informed caricatures, without any hint of the more sophisticated discussions of philosophical issues that can be found in econometric studies like Haavelmo (1944), Vining (1949), Zellner (1971), Leamer (1978), Sims (1980) or Hendry (1980) (I refrain from mentioning papers that appeared after Boland wrote this passage and acknowledge all shortcomings in these papers). It is not easy to take Boland's remark seriously. He briefly discusses the Neyman–Pearson methodology (Boland, 1982, p. 126) but does not analyse its merits or weaknesses.

A final example of methodological interest in econometrics comes from McCloskey who (as already noted) claims that 'no proposition about economic behavior has yet been overturned by econometrics' (McCloskey, 1985, p. 182) (Spanos, 1986, p. 660, makes the same assertion). In Popperian terms, econometrics is a failure. I concur with McCloskey's view and share the critique on the use of significance tests. However, I disagree with the 'rhetorical' cure. The fact that few writers are so well versed in rhetorics as McCloskey herself, combined with my feeling that she does not persuade (me personally, nor the majority of the econom(etr)ics profession), should be sufficient reason to reject the rhetorical approach to methodology on its own terms.

3 Frequency and beliefs

3.1 Frequentist interpretations of probability and econometrics

Von Mises' frequency theory is the most 'objective' basis for probabilistic inference. Von Mises argues that a statistician needs a collective: no collective – no probability. His theory of induction can be summarized by his 'second law of large numbers', which states that if one has a large number of observations, the likelihood function will dominate the prior (see page 37 above). Therefore, we can safely ignore the problem of formulating priors. For econometrics, this has three drawbacks. First, in many cases (in particular, time-series econometrics), the number of observations is rather small. Secondly, the validity of the randomness condition is doubtful. In economics 'stable frequency distributions' rarely exist. Thirdly, even if the number of observations increases, the 'sampling uncertainty' may decrease but the 'specification uncertainty' remains large (Leamer, 1983). As a result of those three considerations, the 'second law of large numbers' cannot be easily invoked. Perhaps not surprisingly, von Mises' theory of inference has not gained a foothold in econometrics.

R. A. Fisher shares the frequentist perspective, without relying on the notion of a 'collective'. His small sample approach based on hypothetical populations should appeal to econometricians. This is one explanation for Fisher's (mostly implicit) popularity in early econometric writings. The likelihood function is one of the basic tools of applied econometricians, it is like a jackknife (Efron, 1986).[2] Like von Mises, Fisher does not want to rely on the specification of prior probability distributions when they are not objectively given. But instead of the 'second law of large numbers', he proposes the problematic fiducial argument. In the econometric literature, it received the same fate as von Mises' theory: fiducial inference is simply neglected. Still, econometricians are tempted to interpret the likelihood as a posterior, and frequently they interpret confidence intervals in the Bayesian way.

Neyman and Pearson, finally, devise a theory of quality control in a context of repeated sampling. They do not aim at inductive inference. Although their approach is helpful in evaluating the qualities of different test statistics, the underlying philosophy is not relevant to econometric inference. Econometricians are not decision makers with explicit loss functions. Their goals are much closer to Fisher's goals of science.

3.2 Frequentists without frequencies

The frequentist interpretation of probability uses the law of large numbers. Keynes ([1921], CW VIII, p. 368) argues that

> the 'law of great numbers' is not at all a good name for the principle which underlies statistical induction. The 'stability of statistical frequencies' would be a much better name for it... But stable frequencies are not very common, and cannot be assumed lightly.

Von Mises, I think, might well agree with this statement.

Moreover, 'stable frequency distributions, as in biology, do not exist in economic variables'. This observation is not made by a critic of econometrics, like Keynes, or an opponent of frequentist econometrics, like Leamer, but by Tjalling Koopmans (1937, p. 58). If von Mises' conditions for collectives are not satisfied, it is not clear how to justify methods of inference that are (explicitly or implicitly) based on them. Perhaps it is by means of a leap of the imagination. For example, one might argue that the 'rigorous notions of probabilities and probability distributions "exist" only in our rational mind, serving us only as a tool for deriving practical statements of the type described above'. This is not de Finetti speaking, but Haavelmo (1944, p. 48).

Koopmans, Haavelmo and many of their followers use the frequentist interpretation of probability as a convenient metaphor. Metaphors are not forbidden in science, the 'random numbers' that Bayesians use in numerical integration are just another metaphor. Some metaphors are more credible than others, though. In the case of the frequentist metaphor, the credibility is limited (in particular, in cases of macro-economic time-series data). This credibility really breaks down if adherents of the metaphor still pretend that their method is 'objective', unlike rival methods such as Bayesian inference.

3.3 The frequentist–Bayes compromise

The debate on the foundations of probability has not resulted in a consensus. However, some efforts to reconcile the frequentist approach with the Bayesian have been made. De Finetti proposed the representation theorem for this purpose (it was discussed in chapter 4, section 3.3). I do not think that this theorem served to settle the debate.

A rather different motivation for introducing objective probability into an overall subjective theory of inference is given by Box (1980). He suggests using frequency methods to formulate a model (this is the specification search) and subsequently suggests using Bayesian methods for purposes of inference and testing. This advice is not rooted in deep philosophical convictions but is more pragmatically inspired by the view that frequentist methods are better suited for diagnostic testing. Indeed, it is not uncommon to encounter a Durbin–Watson statistic in Bayesian analyses. Such statistics reveal whether residuals contain relevant information: such information may be useful in the context of Bayesian inference. What is not valid in this context is to interpret frequentist statistics in frequentist terms.

Good (1988) argues that in a frequentist–Bayes compromise, the Bayes factor against a null hypothesis might be approximated by $1/P\sqrt{N}$ where P is the P-value and N the sample size. In order to avoid Berkson's (1938) paradox, a frequentist statistician should decrease the significance level of a test if the sample size increases. According to Good's rule, a sample of fifty observations would warrant a P-value of 0.14, whereas the conventional 5% significance level is appropriate for sample sizes of 400 observations. This is rarely practised in econometrics. The correlation between P and N in empirical papers in the *Journal of Econometrics* (1973–90) is even positive (see Keuzenkamp and Magnus, 1995, for evidence). This is consistent with the view that econometricians want to measure, not test.

4 Farewell to truth

The alternative responses to Humean scepticism, presented in chapter 1, have different views on the status of truth in scientific inference. Apriorists believe they can obtain 'true' knowledge by thinking. Popperians come 'closer to the truth' by weeding out false propositions. Conventionalists, like Hume himself, and pragmatists like Peirce, do not think that the notion 'truth' is very helpful in the development of knowledge. In the probabilistic response, two schools oppose each other: frequentists, who praise 'truth', and epistemologists, who join Hume and the pragmatists in their farewell to 'truth'.

4.1 Truth and probabilistic inference

The title of this monograph is a play on words, referring to the book of the positivist probability theorist, Richard von Mises: *Probability, Statistics and Truth*. Von Mises argues for a frequentist interpretation of probability: use this for the statistical methods of inference, and apply it in empirical research. This will pave the way to 'truth', where his understanding of truth is a positivist one. Von Mises' 'truth' primarily refers to the 'true' value of the first moment of a distribution (cf. von Mises, [1928] 1981, p. 222). He aims to show that, '[s]tarting from a logically clear concept of probability, based on experience, using arguments which are usually called statistical, we can discover truth in wide domains of human interest' (p. 220).

The frequentist interpretation of probability presumes a state of 'truth', which not only is to be discovered but also enables one to define consistency of estimators and similar notions. Von Mises' probability limits are his 'truth' (but the parameters will always remain stochastic!). Fisher argues that the parameters that one tries to infer are 'true' but unknown constants, characterizing specific probability distributions. If you know *a priori* that a specific distribution is valid, then this way of arguing may be fruitful. And indeed, Fisher has tried to justify the application of specific distributions, in particular the normal distribution, by his theory of experimental design (Fisher, [1935] 1966). Haavelmo is influenced by Fisher's interpretation of 'truth'. Consider, for example, the following statement of Haavelmo (1944, p. 49):

> the question arises as to which probability law should be chosen, in any given case, to represent the 'true' mechanism under which the data considered are being produced. To make this a rational problem of statistical inference we have to start out by an axiom, postulating that every set of observable variables has associated with it one particular 'true,' but unknown, probability law.

The way to interpret this statement is that there is a specific probability distribution, of which the 'true' parameters are unknown. But, unlike Fisher, econometricians do not have a theory of experimental design that validates the choice of a specific distribution. The recent rise of non-parametric inference is a (late) response to this defect, the alternative is to elaborate a theory of specification search (Leamer, 1978).

Few scientists dare to acknowledge that they have problems with 'truth'. De Finetti ([1931] 1989) is one: 'As a boy I began to comprehend that the concept of "truth" is incomprehensible.' The epistemological interpretation of probability has no need for truth. Instead, it emphasizes belief, as for example in Keynes' writings:

> Induction tells us that, on the basis of certain evidence, a certain conclusion is reasonable, not that it is true. If the sun does not rise tomorrow, if Queen Anne still lives, this will not prove that it was foolish or unreasonable of us to have believed the contrary. ([1921]) CW VIII, p. 273)

Keynes is interested in rational belief, not truth. This distinguishes him from Popper, who is after truth. Keynes' perspective is shared by Peirce, as the epigraph to this chapter shows. In his discussion of pragmatism, Peirce writes:

> If your terms 'truth' and 'falsity' are taken in such senses as to be definable in terms of doubt and belief and the course of experience (as for example they would be if you were to define the 'truth' as that ... belief in which belief would tend if it were to tend indefinitely toward absolute fixity), well and good: in that case, you are only talking about doubt and belief. But if by truth and falsity you mean something not definable in terms of doubt and belief in any way, then you are talking of entities of whose existence you can know nothing, and which Ockham's razor would clean shave off. (1958, p. 189)

The only way to interpret 'truth' is the limit of inquiry where a scientific community settles down. But we can never be sure that this settlement is a final one. There is no guarantee that a consensus will emerge, we can only hope for it (see also Hacking, 1990, p. 212).

Peirce subscribes to the growth-of-knowledge school, which believes that science becomes more and more accurate (empirically adequate). This was a mainstream view held by physicists around the end of the nineteenth century. It was even thought that most major discoveries had been made, physics was virtually finished. This reminds us of Keynes ([1921] CW VIII, p. 275), who warns: 'While we depreciate the former probability of beliefs which we no longer hold, we tend, I think, to exaggerate the present degree of certainty of what we still believe.'

4.2 The seduction of objectivity

Objective knowledge is tempting. It is a favourite goal in (neo-) Popperian writing. Objectivity is also a theme that splits the statistics (and econometrics) community. According to Efron (1986, p. 3), objectivity is one of the crucial factors separating scientific thinking from wishful thinking. Efron thinks that, 'by definition', one cannot argue with a subjectivist and, therefore, Efron rejects de Finetti's and Savage's subjectivist interpretation of probability as 'unscientific'. Efron (p. 331) adds that measures of evidence that can be used and understood by the 'scientist in the street' deserve the title 'objective', because they can be directly interpreted by the members of the scientific community. As the Fisherian methods of inference form the most popular language in statistics, they are the objective ones. If we take this argument seriously, then objectivity is what the majority subscribes to. This is like Peirce's conception of truth – with a twist. According to Efron, objectivity is something that pleases and seduces the scientist. If objectivity is what the majority subscribes to, and objectivity is aimed for, this leads to uncritical herd behaviour. Indeed: 'The false idol of objectivity has done great damage to economic science' (Leamer, 1983, p. 36). Moreover, if objectivity is interpreted as a democratic group decision, and the members of the group do not share the same utility (or loss) function, then there may be no rational group decision (Savage, [1954] 1972, p. 172–7). This argument resembles Arrow's impossibility theorem.

The debate on 'objective' versus 'subjective' is not fruitful. In a sense, all methods that are well explained by a researcher are 'objective'. If it can be replicated, it is objective. Inference, in this sense, is usually objective, whether from a frequentist or a Bayesian perspective. Similarly, the data are, both for Bayesians and for non-Bayesians, 'objective'. The real issue is that there is no unique way to get rid of specification uncertainty. The interpretation is always bound to human caprice. If interpretations of different researchers converge, then we might say that the state of knowledge has increased. But this convergence should not be preimposed by a frequentist tyrant.[3]

4.3 Limits to probabilism

A proper analysis of inference with respect to probabilistic hypotheses is of central importance for a sound methodology. Epistemological probability is the tool that serves this analysis. However, cognitive limitations prohibit an ideal use of Bayes' theorem. Scientists are not able to behave like grand-world bookmakers. Even if they face true decision problems, they usually have to construct a small world which may be an improper reduction of the large world. This may lead to surprise and re-specifying the 'small world' (Savage, [1954] 1972). Fully coherent behaviour, such as proposed by de Finetti, is superhuman. De Finetti's claim that the key to every activity of the human mind is Bayes' theorem is untenable. It would make the rational individual a robot (or Turing machine), which may be a correct interpretation of rationality, but the creative individual would be better off being able to escape from the probabilistic straitjacket imposed by the principle of inverse probability. Creativity is the process of changing horizons, moving from one small world to another one. If large world inference is beyond human reach, then creativity (which is by nature hard to model) and incoherent behaviour are indispensable. The possibility to revise plans is typical for human beings. Perhaps the ability to break with coherency is the most precious gift to humanity.

Incoherent behaviour is to sin against the betting approach of Bayesian inference, the one proposed by Ramsey, de Finetti and Savage. However, to sin may be useful, as was already known to the scholastics: 'multae utilitates impedirentur si omnia peccata districte prohiberentur' (Thomas Aquinas, *Summa Theologia*, II. ii, q. 78 i), i.e. much that is useful would be prevented if all sins were strictly prohibited. Reverend Bayes should allow us to sin and disregard the plea for strict coherency. Van Fraassen (1989, p. 176) suggests adopting a liberal version of Bayesian inference in which a certain amount of voluntarism is accepted. He compares this voluntarism with English law, where anything not explicitly forbidden is allowed. Orthodox Bayesianism, on the other hand, is better compared with Prussian law, where anything not explicitly allowed is forbidden (van Fraassen, 1989, p. 171).

Apart from the impossibility of fully coherent behaviour, supporters of the probabilistic approach to inference have to face the problem of specifying prior probability distributions. Keynes ([1921] CW VIII) had already pointed to the problem of obtaining numerical probabilities. Many empirical investigators who have tried to use Bayesian methods for empirical inference have experienced this problem. Jeffreys tried to circumvent this by means of a theory of non-informative prior probability distributions. Although this theory is useful, it is not always satisfactory (for example, when models of different dimension have to be compared). The theory of universal priors is the most elaborate effort to improve Jeffreys' methods. In chapter 5, I showed that this approach runs counter to a logical problem of non-computability, the so-called halting problem. 'True' prior distributions for empirical applications do not generally exist. If sufficient data become available, this problem is relatively unimportant. Furthermore, an investigator may try different prior probability distributions in order to obtain upper and lower limits in probabilistic inference (this approach is proposed by Good and Leamer). Most often, the issue at stake is not whether a variable is 'significant', but whether it remains relevant in various acceptable specifications of the problem.

A third limitation to probabilistic inference is due to reflexivity: self-fulfilling prophecies and other interference of the subject with the object. This undermines the convergence condition in frequentist inference. However, a Bayesian investigator does not have to convert to extreme scepticism. The possibility of instability does not invalidate the rationality of induction: see Keynes' statement about Queen Anne, cited above.

The limitations discussed here suggest that probabilistic inference remains ultimately conjectural. This is nothing to worry about, and it is a view shared by (neo-) Popperians.

4.4 Econometrics and positivism

In chapter 9, I argued that econometrics belongs to the positivist tradition. It is a tradition with different niches, encompassing the views of divergent scientists such as Pearson, Fisher, Jeffreys, von Mises and Friedman. In the recent philosophical literature, positivism has been given a new impulse by van Fraassen (1980). He argues for 'empirically adequate representations' of the data. If this is supplemented by a theory of simplicity, it becomes possible to formalize adequacy and to show how both theory and measurement can be used in probabilistic inference.

The traditional positivist picture of science is one of steady progress. Pearson ([1892] 1911, pp. 96–7), for example, argues that the progress of science lies in the invention of more general formulae that replace more complex but less comprehensive 'laws' (mental shorthand for descriptions of the phenomena). 'The earlier formulae are not necessarily wrong,[4] they are merely replaced by others which in briefer language describe more facts' (p. 97). Jeffreys ([1931] 1957, p. 78), who admires Pearson's *Grammar of Science*, presents a view of science that resembles Pearson's:

> Instead of saying that every event has a cause, we recognize that observations vary and regard scientific method as a procedure for analysing the variation. Our starting point is to consider all variation as random; then successive significance tests warrant the treatment of more and more as predictable, and we explicitly regard the method as one of successive approximation.

I feel sympathetic to this view, although this does not imply that there should be a convergence of opinions (in Peirce's view, this is the same as convergence to truth). This early positivist's view of the growth of knowledge may be too simplistic in general, as in economics the researcher tries to represent a 'moving target', due, for example, to unstable behaviour. Still, in many cases, the target moves slowly enough to enable the investigator to obtain better models (empirically adequate and parsimonious); the analysis of consumer behaviour is a good example.

5 Conclusion

Economic methodology never had much interest in econometrics. This is a pity, as probabilistic inference is one of the strong contenders among the responses to Humean scepticism. Econometrics is, therefore, of particular interest. Analysing the merits and limits of econometrics should be an important activity of methodologists. In this book, I have tried to fill some of the gaps in the methodological literature.

I have argued that econometricians, who search for philosophical roots, should turn to an old tradition that has been declared dead and buried by most economic methodologists: positivism. The works of Pearson, Fisher and Jeffreys deserve more attention than those of the most popular philosopher among economists, Popper. Econometrics is not a quest for truth. Econometric models are tools to be used, not truths to be believed (Theil, 1971, p. vi). And I agree with a warning, expressed by Haavelmo (1944, p. 3), 'it is not to be forgotten that they [our explanations] are all our own artificial inventions in a search for an understanding of real life; they are not hidden truths to be "discovered" '.

Clearly, econometricians have not been able to solve all their problems. In particular, specification uncertainty hampers probabilistic inference. Econometric inference is handicapped by the lack of experimental data (Fisher's context) or large and homogeneous samples (von Mises' context). Neither have econometricians, privately or as a group, well-specified loss functions on which Neyman–Pearson decisions can be based. Hence, the fact that most econometricians still subscribe to the frequentist interpretations of probability is striking. Some explanations can be given. Frequentist tools serve to describe and organize empirical data. They can be used as tools to obtain hypotheses. This is Friedman's interpretation of Tinbergen's work, it is also the view of Box (1980). The next step, inference (estimating parameters, testing or comparing hypotheses, sensitivity analysis), is very often done in an informal, Bayesian way, as if the investigator had used a non-informative prior all along.

Neither the official frequentist methods nor the official Bayesian methods are able to deal adequately with specification uncertainty. A statement of the founder of the probabilistic approach to inference, Laplace (cited in K. Pearson, 1978, p. 657), is particularly appropriate:

> The theory of probability has to deal with considerations so delicate, that it is not surprising that with the same data, two persons will find different results, especially in very complicated questions.

Economic questions deal frequently with complex matter. If complexity is merged with misunderstanding of the foundations and aims of inference, the result will be frustration. This is the fate of econometrics – unless it abandons the quest for truth and returns to its positivist roots.

Notes

1. It is fair to say that the amount of money spent on the making of the atomic bomb is incomparable to the money spent on the development of econometrics: Rhodes (1986) provides an astonishing account of the magnitude of economic costs of this piece of physical engineering.
2. Of course, not all econometricians rely on maximum likelihood. Fisher was opposed by Karl Pearson, who preferred the method of moments. A similar discussion can occasionally be found in econometrics. For example, Sargent (1981, p. 243) prefers the method of moments because it avoids using the likelihood function.
3. See also A. F. M. Smith in his comment to Efron (1986). Note how statisticians use rhetorical methods to make their points. Fisher rejects the Neyman–Pearson methodology as a 'totalitarian' perversion (see Fisher, 1956, p. 7, pp. 100–2; chapter 3, above). Similarly, Smith rejects Efron's interpretation of Fisher's methodology, because '[a]ny approach to scientific inference which seeks to legitimize an answer in response to complex uncertainty is, for me, a totalitarian parody of a would-be rational human learning process' (Smith, in Efron, 1986, p. 10). Rhetorical techniques are used. How successful they are may be questioned.
4. Pearson adds a footnote here, that reads: 'They are what the mathematician would term "first approximations", true when we neglect certain small quantities. In Nature it often happens that we do not observe the existence of these small quantities until we have long had the "first approximation" as our standard of comparison. Then we need a widening, not a rejection of "natural law".'

Personalia

Thomas Bayes (1702–61). English reverend whose posthumous *Essay Towards Solving a Problem in the Doctrine of Chance* (1763) laid the foundation for Bayesian inference. See also Laplace.

Emile Borel (1871–1956). French mathematician who contributed to Bayesian probability theory and measure theory.

Arthur Lyon Bowley (1869–1957). Professor of statistics at the London School of Economics, founding member and later president of the Econometric Society. Was sympathetic to Bayesian inference.

Rudolf Carnap (1891–1970), philosopher, founder of logical positivism. Carnap studied in Vienna. Professor of philosophy in Chicago from 1936 to 1952. Subsequently, he succeeded Reichenbach at the University of California at Los Angeles, where he held a chair until 1961. Carnap belongs to the most important opponents of Popper in epistemology through his support for inductive inference.

Harald Cramér (1893–1985). Swedish probability theorist who contributed to central limit theory. His probability theory is closely related to the theory of von Mises (Cramér, 1955, p. 21). Cramér's (1946) book on mathematical statistics was the standard reference book on probability theory for econometricians of the postwar decade. One of his students was Herman Wold.

Bruno de Finetti (1906–85). Italian probability theorist. One of the most radical twentieth-century contributors to the subjective approach to probability theory. He invented the notion of exchangeability.

Pyrrho of Elis (*c.* 365–275 BC), the first and most radical Sceptic, his philosophy came to be known as Pyrrhonism. He dismissed the search for truth as a vain endeavour. See also Sextus Empiricus and David Hume.

Sextus Empiricus (*c.* second century AD). Greek philosopher, popularizer of Pyrrhonism (in *Outlines of Pyrrhonism* and *Against the Dogmatists*). His work was re-published in 1562 and had much influence on the philosophy of, e.g., Descartes and Hume.

Ezekiel, Mordecai (1899–1974). Agricultural economist, who wrote an influential textbook on econometrics-*avant-la-lettre*.

Ronald Aylmer Fisher (1890–1962). English statistician and geneticist, one of the greatest probability theorists of the twentieth century. He invented the maximum likelihood method, provided the foundations for the t-statistic, invented experimenal design based on randomization and contributed many statistical concepts. Sadly, philosophers and methodologists rarely read his work.

Gottlob Frege (1848–1925). German mathematician and logician. Influenced and inspired logical positivism via Russell and Wittgenstein.

Ragnar Frisch (1895–1973). Norwegian economist who invented the word 'econometrics' and helped to found the Econometric Society. His main econometric interest was related to systems of equations, for which he invented the 'bunch map' technique. Kalman may be regarded as one of his heirs.

Francis Galton (1822–1911). English eugenicist and statistician, known for his analysis of correlation and 'regression'. Among his less-known contributions are a statistical test of the efficiency of prayer.

Carl Friedrich Gauss (1777–1855). German mathematician and probability theorist. He provided the probabilistic context for least squares approximation. Gauss is a contender for being named as inventor of the method of least squares, but this claim seems to be unwarranted (see Legendre).

William Gosset (1876–1937). English statistician who contributed to small sample analysis while solving practical problems at Guinness. In particular known for the Student t-distribution (published in 1908 under the pseudonym Student).

Trygve Haavelmo (1911–). Norwegian econometrician, student of Frisch. During the 1943–47 period, he was research associate at the Cowles Commission. While in the US, he had contacts with Neyman, Wald and many other statisticians. Haavelmo (1944) strongly influenced the course of econometrics in the postwar years, with its emphasis on simultaneous equations bias and formal Neyman–Pearson analysis.

David Hume (1711–76). Scottish philosopher and economist. Known for his philosophical scepticism and empiricism (partly inspired by the writings of Sextus Empiricus). Hume claims that it is not possible to verify the truth of (causal) relations.

Harold Jeffreys (1891–1989). British scientist and probability theorist. He made important contributions to make Bayesian inference of practical use, in particular by elaborating non-informative prior probability distributions and the method of posterior odds.

Immanuel Kant (1724–1804). German philosopher, known for the 'synthetic truth *a priori*'.

John Maynard Keynes (1883–1946). English probability theorist who also has some reputation as an economist. His *Treatise on Probability* counts as one of the most important works on the theory of scientific induction of the twentieth century.

John Neville Keynes (1852–1949). English economist who contributed John Maynard Keynes to society.

Andrei Kolmogorov (1903–87). Russian mathematician and probability theorist. Known for his formal axiomatic approach to probability.

Tjalling Koopmans (1910–85). Dutch (later American) physicist who turned to econometrics and economic theory. He wrote his dissertation on econometrics under supervision of Tinbergen (Koopmans, 1937). After a short stay at the League of Nations, he moved to the USA in 1940, joining the Cowles Commission in 1944 where he elaborated the formal approach to econometrics.

Imre Lakatos (1922–74). Hungarian philosopher who refined the Popperian methodology of scientific research programmes.

Pierre Simon de Laplace (1749–1827). Founder of probability theory who rediscovered 'Bayes' Theorem'. The contributions of Laplace, among others, in the *Théorie Analytique des Probabilités* (1812), go far beyond those of Bayes. Despite his interest in probability theory, his philosophy was strictly deterministic.

Adrien Marie Legendre (1752–1833). Successor to Laplace at the École Militaire and École Normale. He invented the method of least squares, although some (e.g. Spanos, 1986, p. 448; see also p. 253) attribute priority to Gauss. The priority dispute between Gauss and Legendre is described in Stigler (1986, pp. 145–6).

Ernst Mach (1838–1916). 'Father of logical positivism', physicist.

Ludwig von Mises (1881–1973). Brother of Richard. 'Austrian' economist who opposed the quantitative and statistical approach to economics.

Richard von Mises (1883–1953). Physicist and probability theorist. Born in Russia. Von Mises studied in Vienna where he became a disciple of the positivist Mach. Later, when he held a chair in Berlin, he regularly attended Moritz Schlick's meetings of the *Wiener Kreis*. At these meetings, he met Popper (see Popper, 1976), but von Mises neither became a logical positivist nor a Popperian falsificationist. He disagrees with Carnap on inductive probability (e.g. von Mises, [1928] 1981, p. 96).

Henry Ludwell Moore (1869–1958). Pioneer of econometrics, student of Karl Pearson. Moore was involved in statistical analysis of demand and business cycles.

Jerzy Neyman (1894–1981). Polish probability theorist who emigrated in 1938 to the USA. With Egon Pearson, he developed a behaviourist theory of hypothesis testing and quality control. He co-translated the first German edition of von Mises ([1928] 1957).

William of Ockham (c. 1285-1349). Oxford Franciscan scholastic philosopher, known for his 'razor'.

Egon Pearson (1896–1980). English statistician, son of Karl. Collaborator of Jerzy Neyman with whom he developed the theory of hypothesis testing.

Karl Pearson (1857–1936). English statistician, who developed the theory of distributions and invented the *p*-value.

Charles Peirce (1839–1914). American philosopher, contributed to the philosophy of 'pragmaticm'.

Henri Poincaré (1854–1912). French mathematician, astronomer and philosopher of science. His *La Science et l'hypothèse* contributes to the positivist philosophy of science and probability theory.

Karl Popper (1902-94). Over-rated English (Austrian born) philosopher of science.

Adolphe Quetelet (1796–1874). Belgian sociologist and statistician. Applied statistics to a wide variety of natural and social phenomena (note, e.g., his concept of the 'average man'). Quetelet met Laplace and Poisson during a visit to Paris in 1823. Since then, he became interested in applying statistics to social phenomena (see Keynes, [1921] CW VIII, p. 366).

Frank Ramsey (1903–30). English mathematician, economist, probability theorist and philosopher. He provided a formal foundation of the personalist theory of probability. He translated Wittgenstein's *Tractatus Logico-Philosophicus* into English.

Hans Reichenbach (1891–1953). Logical positivist who worked on a frequentist philosophy of probability. Held professorships at Berlin and Istanbul and later moved to UCLA.

Leonard J. ('Jimmie') Savage (1917–71). American statistician who elaborated the personalist or subjective theory of probability. He combined subjective probability with utility theory.

Moritz Schlick (1882–1936). German philosopher and founder of the *Wiener Kreis*. This group included Rudolph Carnap, Otto Neurath, Kurt Gödel, Philip Frank and Hans Hahn.

Henry Schultz (1893–1938). Russian emigrant to the United States who studied in London (LSE and Galton Laboratory of University College) under Bowley and Karl Pearson. He was influenced by Henry L. Moore during his later study at Columbia University. Schultz made a path breaking contribution to the statistical analysis of consumer demand.

'Student'. See Gosset.

Jan Tinbergen (1903–94). Dutch physicist and econometrician. He worked at the League of Nations (1936-8) for which we wrote a two volume work on business cycles, which aimed at quantifying economic theories. The first volume was critically reviewed by J. M. Keynes.

Abraham Wald (1902–50). Mathematician and probability theorist. He studied in Vienna between 1927 and 1931. He probably met von Mises in Vienna at Karl Menger's Mathematisches Colloquium (see Popper, 1976). Wald left Austria in 1938 and found shelter at the Cowles Commission. In 1939, he went to Columbia University to work with the statistician Harold Hotelling.

Norbert Wiener (1884–1964). Information theorist who worked at MIT.

Ludwig Wittgenstein (1889–1951). Austrian philosopher.

George Udny Yule (1871–1951). English statistician. Obtained a 'demonstratorship' for Karl Pearson in 1893 and became lecturer at Cambridge in 1912. Yule is the inventor of the multiple correlation technique (Stigler, 1986, p. 354). One of Yule's most important accomplishments is the connection of the method of least squares to Galton's regression model of conditional expectation (pp. 285 and 295). Another important contribution is the analysis of 'nonsense correlation'.

References

Aigner, Dennis J. and Arnold Zellner (1988), Causality: Editor's introduction, *Journal of Econometrics* 39, 1–5.

Amemiya, Takeshi (1985), *Advanced Econometrics*, Harvard University Press, Cambridge, MA.

Angrist, Joshua D. (1990), Lifetime earnings and the Vietnam era draft lottery: evidence from social security administrative records, *American Economic Review* 80, 313–36.

Backhouse, Roger E. (1994, ed.), *New Directions in Economic Methodology*, Routledge, London.

Baird, Davis (1988), Significance tests, history and logic, in Samuel Kotz and Norman L. Johnson, eds. (1988), *Encyclopedia of Statistical Sciences* 8, John Wiley, New York.

Barnard, George A. (1992), Review of: *Statistical inference and analysis: selected correspondence of R. A. Fisher*, *Statistical Science* 7, 5–12.

Barnett, Vic (1973), *Comparative Statistical Inference*, John Wiley, London.

Barten, Anton P. (1967), Evidence on the Slutsky conditions for demand equations, *Review of Economics and Statistics* 49, 77–84.

(1969), Maximum likelihood estimation of a complete system of demand equations, *European Economic Review* 1, 7–73.

(1977), The systems of consumer demand functions approach: a review, *Econometrica* 45, 23–51.

Basmann, R. L. (1963), The causal interpretation of non-triangular systems of economic relations, *Econometrica* 31, 439–48.

Bentham, Jeremy (1789), *An Introduction to the Principles of Morals and Legislation*, Oxford.

Berger, James O. ([1980] 1985), *Statistical Decision Theory and Bayesian Analysis*, Springer Verlag, New York.

Berkson, J. (1938), Some difficulties of interpretation encountered in the application of the chi-squared test, *Journal of the American Statistical Association* 33, 526–42.

Binmore, Ken (1991), Modeling rational players II, *Economics and Philosophy* 4, 9–55.

Blaug, Mark (1980), *The Methodology of Economics or How Economists Explain*, Cambridge University Press.

([1962] 1985), *Economic Theory in Retrospect*, Cambridge University Press.

Blundell, Richard (1988), Consumer behaviour: theory and empirical evidence – a survey, *Economic Journal* 98, 16–65.

Boehner, Philotheus (1989), *William of Ockham, philosophical writings, a selection.*

Boland, Lawrence A. (1982), *The Foundations of Economic Method*, Allen & Unwin, London.

(1989), *The Methodology of Economic Model Building*, Routledge, London.

Bowden, Roger J. (1989), *Statistical Games and Human Affairs, the View from Within*, Cambridge University Press.

Bowley, Arthur L. ([1901] 1937), *Elements of Statistics*, King & Son, Westminster.

Box, George E. P. (1980), Sampling and Bayes inference in scientific modelling and robustness, *Journal of the Royal Statistical Society A* 143, 383–430 (with discussion).

Brown, A. and Angus Deaton (1972), Surveys in applied economics: models of consumer behaviour, *Economic Journal* 82, 1145–236.

Brown, Bryan W. and Mary Beth Walker (1989), The random utility hypothesis and inference in demand systems, *Econometrica* 57, 815–29.

Brown, James N. and Robert W. Rosenthal (1990), Testing the minimax hypothesis: a re-examination of O'Neill's game experiment, *Econometrica* 58, 1065–81.

Bunge, Mario ([1959] 1979), *Causality and Modern Science*, New York.

Burns, Arthur F. and Wesley C. Mitchell (1946), *Measuring Business Cycles*, National Bureau of Economic Research, New York.

Byron, R. P. (1970), A simple method for estimating demand systems under separable utility assumptions, *Review of Economic Studies* 37, 261–74.

Caldwell, Bruce J. (1982), *Beyond Positivism, Economic Methodology in the Twentieth Century*, Allen & Unwin, London.

(1991), Clarifying Popper, *Journal of Economic Literature* 29, 1–33.

Campbell, John Y. (1987), Does saving anticipate declining labor income? An alternative test of the permanent income theory, *Econometrica* 55, 1249–73.

Carabelli, Anna (1985), Keynes on cause, chance and possibility, in Lawson and Pesaran, eds. (1985), 151–80.

Carnap, Rudolf (1950), *Logical Foundations of Probability*, University of Chicago Press, Chicago, IL.

(1952), *The Continuum of Inductive Methods*, University of Chicago Press, Chicago, IL.

(1963), *Intellectual autobiography*, in Schilpp (1963), 3–84.

Carnap, Rudolf and Richard Jeffrey, eds. (1971), *Studies in Inductive Logic and Probability*, University of California Press, Berkeley, CA.

Cartwright, Nancy (1989), *Nature's Capacities and Their Measurement*, Clarendon Press, Oxford.

Chambers, Marcus J. (1990), Forecasting with demand systems, a comparative study, *Journal of Econometrics* 44, 363–76.

Chow, Gregory C. (1983), *Econometrics*, McGraw-Hill, Auckland.

Christensen, Laurits R., Dale W. Jorgenson and Lawrence J. Lau (1975), Transcendental logarithmic utility functions, *American Economic Review* 65, 367–83.

Chung, Kai Lai (1974), *A Course in Probability Theory*, Academic Press, New York.

Cohen, L. Jonathan (1989), *An Introduction to the Philosophy of Induction and Probability*, Clarendon Press, Oxford.

Cohen, L. Jonathan and Richard S. Westfall (1995), *Newton, Texts, Backgrounds, Commentaries*, Norton, New York.

Cox, David R. (1961), Tests of separate families of hypotheses, *Proceedings of the Fourth Berkeley Symposium on Mathematical Statistics and Probability* **1**, University of California Press, Berkeley, CA, 105–23.

(1962), Further results on tests of separate families of hypotheses, *Journal of the Royal Statistical Society B* 24, 406–24.

(1992), Causality, some statistical aspects, *Journal of the Royal Statistical Society A* 155, 291–301.

Cox, David R. and D. V. Hinkley (1974), *Theoretical Statistics*, Chapman and Hall, London.

Cramér, Harald (1946), *Mathematical Methods of Statistics*, Princeton University Press, Princeton, NJ.

(1955), *The Elements of Probability Theory and some of its Applications*, John Wiley, New York.

Cramer, J. S. (1987), Mean and variance of R^2 in small and moderate samples, *Journal of Econometrics* 35, 253–66.

Cross, Rod (1982), The Duhem–Quine thesis, Lakatos and the appraisal of theories in macroeconomics, *Economic Journal* 92, 320–40.

Crum, William L., Alson C. Patton and Arthur R. Tebbutt (1938), *Introduction to Economic Statistics*, McGraw-Hill, New York.

Darnell, Adrian C. and J. Lynne Evans (1990), *The Limits of Econometrics*, Edward Elgar, Aldershot, Hants.

Davenant, C. (1699), *An Essay Upon the Probable Methods of Making a People Gainers in the Balance of Trade*, London.

Davis, Harold T. (1941), *Theory of Econometrics*, Principia Press, Bloomington, IN.

Davis, John B. (1994), *Keynes's Philosophical Development*, Cambridge Universisty Press.

Dawid, A. P. M. Stone and J. V. Zidek (1973), Marginalization paradoxes in Bayesian and structural inference, *Journal of the Royal Statistical Society B* 35, 189–233.

Deaton, Angus (1974), The analysis of consumer demand in the United Kingdom, 1900–1970, *Econometrica* 42, 341–67.

(1978), Specification and testing in applied demand analysis, *Economic Journal* 88, 524–36.

(1990), Price elasticities from survey data, extensions and Indonesian results, *Journal of Econometrics* 44, 281–309.

Deaton, Angus and John Muellbauer (1980a), An Almost Ideal Demand System, *American Economic Review* 70, 312–36.

(1980b), *Economics and Consumer Behavior*, Cambridge University Press.

De Marchi, Neil, ed. (1988), *The Popperian Legacy in Economics*, Cambridge University Press.

De Marchi, Neil and Mark Blaug, eds. (1991), *Appraising Economic Theories, Studies in the Methodology of Research Programmes*, Edward Elgar, Hants.

Derkse, W. (1992), *On Simplicity and Elegance: an Essay in Intellectual History*, Eburon, Delft.

Dharmapala, Dhammika and Michael McAleer (1995), Prediction and accommodation in econometric modelling, *Environmetrics* 6, 551–6.

Diewert, W. E. (1971), An application of the Shephard duality theorem: a generalized Leontief production function, *Journal of Political Economy* 79, 481–507.

(1974), Intertemporal consumer theory and the demand for durables, *Econometrica* 42, 497–516.

Doan, Thomas A. (1988), *User's Manual RATS 3.00*, VAR Econometrics, Evanston, MN.

Doan, Thomas, Robert Litterman and Christopher Sims (1984), Forecasting and conditional projection using realistic prior distributions, *Econometric Reviews* 3, 1–100.

Earman, John, ed. (1983), *Testing Scientific Theories,* Minnesota Studies in the Philosophy of Science 10, University of Minnesota Press, Minneapolis, MN.

Eatwell, John, Murray Milgate and Peter Newman, eds. (1987), *The New Palgrave: a Dictionary of Economics,* Macmillan, London.

Edwards, A. W. F. ([1972] 1992), *Likelihood*, The Johns Hopkin University Press, Baltimore, MD.

Efron, Brad (1986), Why isn't everyone a Bayesian? (with discussion), *American Statistician* 40, 1–11 and 330–1.

Ehrenberg, A. S. C. and J. A. Bound (1993), Predictability and prediction, *Journal of the Royal Statistical Society A* 156, 167–206.

Ehrlich, Isaac and Zhiqiang Liu (1999), Sensitivity analysis of the deterrence hypothesis: let's keep the econ in econometrics, *Journal of Law and Economics*.

Eichenbaum, Martin (1995), Some comments on the role of econometrics in economic theory, *Economic Journal* 105, 1609–21.

Ellis, Robert Leslie (1843), On the foundations of the theory of probability, *Transactions of the Cambridge Philosophical Society* 8.

Engle, Robert F. (1984), Wald, Likelihood Ratio, and Lagrange Multiplier tests in econometrics, in Z. Griliches and M. D. Intriligator, eds. (1984), *Handbook of Econometrics* 2, North Holland, Amsterdam, 775–826.

Epstein, Roy (1987), *A History of Econometrics*, Elseviers Science Publishers, Amsterdam.
Ezekiel, Mordecai (1930), *Methods of Correlation Analysis*, Wiley, New York.
(1941), *Methods of Correlation Analysis*, second edition, Wiley, New York.
Ferber, Robert and Werner Z. Hirsch (1982), *Social Experimentation and Economic Policy*, Cambridge University Press.
Feyerabend, Paul (1975), *Against Method*, Verso, London.
(1987), Trivializing knowledge: Comments on Popper's excursions in philosophy, in his *Farewell to Reason*, Verso, London, 162–91.
Feynman, Richard P. (1965), *The Character of Physical Law*, The MIT Press, Cambridge, MA.
(1985), *Surely You're Joking, Mr. Feynman*, Bantam, New York.
Finetti, Bruno de ([1931] 1989), Probabilism, a critical essay on the theory of probability and on the value of science, English translation in: *Erkenntnis* 31, 169–223.
(1974), *Theory of Probability*, vol. I, John Wiley, New York.
(1975), *Theory of Probability*, vol. II, John Wiley, New York.
Fisher, Ronald A. (1915), Frequency distribution of the values of the correlation coefficient in samples from an indefinitely large population, *Biometrika* 10, 507–21.
(1922), On the mathematical foundations of theoretical statistics, *Philosophical Transactions of the Royal Society A*, 222, 309–68.
([1925] 1973), *Statistical Methods for Research Workers*, Hafner Publishing Company, New York.
([1935] 1966), *The Design of Experiments*, Hafner Publishing Company, New York.
(1955), Statistical methods and scientific induction, *Journal of the Royal Statistical Society B* 17, 69–78.
(1956), *Statistical Methods and Scientific Inference*, Oliver and Boyd, London.
Fisher Box, Joan (1978), *R. A. Fisher, The Life of a Scientist*, John Wiley & Sons, New York.
Florens, Jean-Pierre, Michel Mouchart and Jean-Marie Rolin (1990), *Elements of Bayesian Statistics*, Marcel Dekker, New York.
Forster, Malcolm R. (1995), Bayes and bust: simplicity as a problem for a probabilist's approach to confirmation, *British Journal for the Philosophy of Science* 46, 399–424.
(2000) The new science of simplicity, in Keuzenkamp, McAleer and Zellner (2000).
Fox, Karl A. (1989), Agricultural economists in the econometric revolution: institutional background, literature and leading figures, *Oxford Economic Papers* 41, 53–70.
Fraassen, Bas C. van (1980), *The Scientific Image*, Clarendon Press, Oxford.
(1989), *Laws and Symmetry*, Clarendon Press, Oxford.
Frazer, W. J. and L. A. Boland (1983), An essay on the foundations of Friedman's methodology, *American Economic Review* 73, 129–44.

Friedman, Milton (1940), Review of Tinbergen's *Statistical Testing of Business Cycle Theories, II*, *American Economic Review* 30, 657–61.

(1953), *Essays in Positive Economics*, University of Chicago Press, Chicago, IL

(1992), Do old fallacies ever die? *Journal of Economic Literature* 30, 2129–32.

Friedman, Milton and Rose D. Friedman (1998), *Two Lucky People*, University of Chicago Press, Chicago, IL.

Friedman, Milton and Anna J. Schwartz (1963), *A Monetary History of the United States 1867–1960*, Princeton University Press, Princeton, NJ.

(1991), Alternative approaches to analyzing economic data, *American Economic Review* 81, 39–49.

Frisch, Ragnar (1934), *Statistical Confluence Analysis by Means of Complete Regression Systems*, Institute of Economics, Oslo University.

Gans, Joshua S. and George B. Shepherd (1994), How are the Mighty fallen: rejected classic articles by leading economists, *Journal of Economic Perspectives* 8, 165–79.

Gauss, Carl Friedrich ([1809] 1963), *Theory of Motion of the Heavenly Bodies Moving About the Sun in Conic Sections*, translated from Latin by C. H. Davis, Dover, New York.

Gaver, Kenneth M. and Martin S. Geisel (1974), Discriminating among alternative models: Bayesian and non-Bayesian methods, in Zarembka (1974), 49–77.

Geweke, John (1982), Causality, exogeneity and inference, in Werner Hildenbrand, ed. (1982), *Advances in Econometrics*, Cambridge University Press, 209–35.

Giere, Ronald N. (1983), Testing theoretical hypotheses, in Earman (1983), 269–98.

(1988), *Explaining Science, a Cognitive Approach*, University of Chicago Press, Chicago, IL.

Gilbert, Chris (1991a), Do economists test theories? Demand analysis and consumption analysis as tests of theories of economic methodology, in De Marchi and Blaug (1991), 137–68.

(1991b), Richard Stone, demand theory and the emergence of modern econometrics, *Economic Journal* 101, 288–302.

Godfrey, L. G. (1989), *Misspecification Tests in Econometrics*, Cambridge University Press.

Goldberger, Arthur S. (1964), *Econometric Theory*, Wiley, New York.

(1998), *Introductory Econometrics*, Harvard University Press, Cambridge, MA.

Good, Irving J. (1965), *The Estimation of Probabilities: an Essay on Modern Bayesian Methods*, The MIT Press, Cambridge, MA.

(1968), Corroboration, explanation, evolving probability, simplicity and a sharpened razor, *British Journal for the Philosophy of Science* 19, 123–43.

(1973), The probabilistic explication of evidence, surprise, causality, explanation, and utility, in V. P. Godambe and D. A. Sprott, eds. (1973), *Foundations of Statistical Inference*, Holt, Rinehart and Winston, Toronto.

(1988), The interface between statistics and philosophy of science (with comments), *Statistical Science* 3, 386–412.

Granger, Clive W. J. (1969), Investigating causal relations by econometric models and cross-spectral methods, *Econometrica* 37, 424–38.

Granger, Clive W. J. and P. Newbold (1974), Spurious regressions in econometrics, *Journal of Econometrics* 2, 111–20.

Granger, Clive and Harald Uhlig (1990), Reasonable Extreme Bounds, *Journal of Econometrics* 44, 159–70.

Greene, William H. (1993), *Econometric Analysis*, second edition, Macmillan, New York.

Haavelmo, Trygve (1944), The probability approach in econometrics, supplement to *Econometrica* 12, July.

Hacking, Ian (1975), *The Emergence of Probability, a Philosophical Study of Early Ideas About Probability, Induction and Statistical Inference*, Cambridge University Press, Cambridge.

(1983), *Representing and Intervening. Introductory Topics in the Philosophy of Natural Science*, Cambridge University Press.

(1990), *The Taming of Chance*, Cambridge University Press.

Hahn, Frank (1985), In praise of economic theory, in his *Money, Growth and Stability*, Basil Blackwell, Oxford, 10–30.

(1992), Answer to Backhouse: Yes, *Royal Economic Society Newsletter*, 78, July.

Hamminga, Bert (1983), *Neoclassical Theory Structure and Theory Development*, Springer, Berlin.

Hands, D. Wade (1991), The problem of excess content: economics, novelty and a long Popperian tale, in De Marchi and Blaug (1991), 58–75.

Hansen, Lars P. and James J. Heckman (1996), The empirical foundations of calibration, *Journal of Economic Perspectives* 10, 87–104.

Hansen, Lars P. and Kenneth J. Singleton (1985), Stochastic consumption, risk aversion and the temporal behaviour of asset returns, *Journal of Political Economy* 91, 249–65.

Hausman, Daniel (1988), An appraisal of Popperian methodology, in De Marchi (1988), 65–85.

(1989), Economic methodology in a nutshell, *Journal of Economic Perspectives* 3, 115–27.

Hayek, Friedrich von (1989), The pretence of knowledge, Nobel Memorial Lecture, 11 December 1974, *American Economic Review* 79, 3–7.

Heckman, James J. (1992), Haavelmo and the birth of modern econometrics, *Journal of Economic Literature* 30, 876–86.

Hendry, David F. (1980), Econometrics: alchemy or science?, *Economica* 47, 387–406.

(1992), Assessing empirical evidence in macroeconometrics with an application to consumers' expenditure in France, in Vercelli and Dimitri (1992), 363–92.

(1993), *Econometrics, Alchemy or Science? Essays in Econometric Methodology*, Blackwell, Oxford.

Hendry, David F. and Neil R. Ericsson (1991), An econometric analysis of UK money demand in *Monetary Trends in the United States and the United Kingdom* by Milton Friedman and Anna J. Schwartz, *American Economic Review* 81, 8–38.

Hendry, David F. and Mary S. Morgan (1995), *The Foundations of Econometric Analysis*, Cambridge University Press.

Hendry, David F., Edward E. Leamer and Dale J. Poirier (1990), The ET dialogue: a conversation on econometric methodology, *Econometric Theory* 6, 171–261.

Hicks, John R. ([1939] 1946), *Value and Capital*, Oxford University Press.

Hildreth, Clifford (1986), *The Cowles Commission in Chicago*, 1939–1955, Springer Verlag, Berlin.

Hill, Carter, William Griffith and George Judge, *Undergraduate Econometrics*, Wiley, New York, 1997.

Hogarth, Robin M. and Melvin W. Reder, eds. (1986), *Rational Choice*, University of Chicago Press, Chicago, IL.

Howson, Colin (1988a), On the consistency of Jeffreys' Simplicity Postulate, and its role in Bayesian inference, *The Philosophical Quarterly* 38, 68–83.

(1988b), Accommodation, prediction and Bayesian confirmation theory, *PSA* 2, 381–92.

Howson, Colin and Peter Urbach (1989), *Scientific Reasoning: the Bayesian Approach*, Open Court, La Salle, IL.

Hume, David ([1739] 1962), *A Treatise of Human Nature, Being an Attempt to Introduce the Experimental Method of Reasoning Into Moral Subjects, Book 1: Of the Understanding*, Fontana Library, London.

([1748] 1977), *An Enquiry Concerning Human Understanding*, ed. Eric Steinberg, Hackett Publishing Company, Indianapolis, IN.

Hutchison, Terence W. (1981), *The Politics and Philosophy of Economics*, New York University Press, New York.

Jaynes, Edwin T. (1973), The well-posed problem, *Foundations of Physics* 3, 477–93.

(1978), Where do we stand on maximum entropy?, in Jaynes (1989), 210–314.

(1979), Concentrations of distributions at entropy maxima, *mimeo*, reprinted in: Jaynes (1989), 315–36.

(1980), Marginalization and prior probabilities, in Zellner (1980), *Bayesian Analysis in Econometrics and Statistics*, North Holland, Amsterdam, 43–78.

(1989), *Papers on Probability, Statistics and Statistical Physics*, Kluwer, Dordrecht.

Jeffreys, Harold ([1931] 1957), *Scientific Inference*, Cambridge University Press.

([1939] 1961), *Theory of Probability*, Clarendon Press, Oxford.

Johansen, L. (1981), Suggestions towards freeing systems of demand equations from a strait-jacket, in: A. Deaton, ed. (1981), *Essays in the Theory and Measurement of Consumer Behaviour in Honour of Sir Richard Stone*, Cambridge University Press, 31–54.

Johnston, J. ([1963] 1984), *Econometric Methods*, McGraw-Hill, Singapore.

Judge, George G., W. E. Griffiths, R. Carter Hill, Helmut Lütkepohl and Tsoung-Chao Lee ([1980] 1985), *The Theory and Practice of Econometrics*, John Wiley, New York.

Kagel, John H. and Alvin E. Roth (1995), *Handbook of Experimental Economics*, Princeton University Press, Princeton, NJ.

Kemeny, John G. (1953), The use of simplicity in induction, *Philosophical Review* 62, 391–408.

(1963), Carnap's theory of probability and induction, in: Schilpp (1963, ed.), 711–38

Keuzenkamp, Hugo A. (1994), What if an idealization is problematic? The case of the homogeneity condition in consumer demand, *Poznan Studies in the Philosophy of Science* 38, 243–54.

(1995), The econometrics of the Holy Grail, *Journal of Economic Surveys* 9, 233–48.

Keuzenkamp, Hugo A. and Anton P. Barten (1995), Rejection without falsification, on the history of testing the homogeneity condition in the theory of consumer demand, *Journal of Econometrics* 67, 103–28.

Keuzenkamp, Hugo A. and Jan Magnus (1995), On tests and significance in econometrics, *Journal of Econometrics* 67, 5–24.

Keuzenkamp, Hugo A. and Michael McAleer (1995), Simplicity, scientific inference and econometric modelling, *Economic Journal* 105, 1–21.

(1997), The complexity of simplicity, *Mathematics and Computers in Simulation,* 1403, 1–9.

(1999), Simplicity, accommodation and prediction in economics, unpublished.

Keuzenkamp, Hugo A., Michael McAleer and Arnold Zellner, eds. (2000), *Simplicity and Econometrics*, Cambridge University Press.

Keynes, John Maynard ([1921] CW VIII), *A Treatise on Probability*, St Martin's Press, New York.

([1936] CW VII), *The General Theory of Employment, Interest and Money*, Macmillan, London.

(1939), Professor Tinbergen's method, *Economic Journal* 49, 558–68, reprinted in: Keynes (1973), 306–18.

(1940), Comment, *Economic Journal* 50, reprinted in: Keynes (1973), 318–21.

([1933] CW X), *Essays in Biography*, Macmillan and Cambridge University Press, London.

(1973, CW XIV), *The General Theory and After, Part II, Defence and Development*, Macmillan St Martin's Press, London.

(1983, CW XI), *Economic Articles and Correspondence: Academic*, Macmillan and Cambridge University Press, London.

Kirzner, Israel M. (1976), On the method of Austrian economics, in Edwin G. Dolan, ed. (1976), *The Foundations of Modern Austrian Economics*, Sheed & Ward, Inc., Kansas City, KS, 40–51.

Klant, J. J. (1979), *Spelregels Voor Economen* (translation: *The Rules of the Game*, Cambridge University Press), Stenferd Kroese, Leiden.

(1990), Refutability, *Methodus* 2, 6–8.

Kolmogoroff, A. (Kolmogorov) (1933), *Grundbegriffe der Wahrscheinlichkeitsrechnung, Ergebnisse der Mathematik und ihrer Grenzgebiet*, II, Verlag von Julius Springer, Berlin.

Koopmans, Tjalling (1937), *Linear Regression Analysis of Economic Time Series*, De Erven F. Bohn N.V., Haarlem.

(1941), The logic of econometric business-cycle research, *Journal of Political Economy* 49, 157–81.

(1947), Measurement without theory, *Review of Economic Statistics* 29, 161–72.

(1949), A reply [to R. Vining, Koopmans on the choice of variables to be studied and of methods of measurement], *Review of Economic Statistics* 31, 86–91.

(1950, ed.), *Statistical Inference in Dynamic Economic Models*, Cowles Commission Monograph 10, Wiley, New York.

Krüger, Lorenz (1987), The slow rise of probabilism, philosophical arguments in the Nineteenth Century, in Krüger, Daston and Heidelberger (1987), 59–89.

Krüger, Lorenz, Lorraine J. Daston and Michael Heidelberger (1987), *The Probabilistic Revolution,* Volume I: *Ideas in History*, The MIT Press, Cambridge, MA.

Kuhn, Thomas S. (1962), *The Structure of Scientific Revolutions*, University of Chicago Press, Chicago, IL.

Kyburg, Henry E. and Howard E. Smokler ([1964] 1980), *Studies in Subjective Probability*, Krieger Publishing Company, New York.

Kydland, Fynn E. and Edward C. Prescott (1994), The computational experiment: an econometric tool, *Federal Reserve Bank of Cleveland Working Paper* 9420.

(1996) The computational experiment: an econometric tool, *Journal of Economic Perspectives*, Winter.

Laitinen, Kenneth (1978), Why is demand homogeneity so often rejected? *Economic Letters* 1, 187–91.

Lakatos, Imre (1970), Falsification and the methodology of scientific research programmes, in: Lakatos and Musgrave (1970), *Criticism and the Growth of Knowledge*, Cambridge University Press, 91–196.

(1978), *Philosophical Papers* Volume I: *The Methodology of Scientific Research Programmes*, Cambridge University Press.

Langley, Pat, Herbert A. Simon, Gary L. Bradshaw and Jan M. Zytkow (1987), *Scientific Discovery, Computational Explorations of the Creative Process*, The MIT Press, Cambridge, MA.

Laplace, Pierre Simon de ([1814] 1951), *Essai Philosophique sur les Probabilités*; English translation: *A Philosophical Essay on Probabilities*, Dover, New York.

Lawson, Tony (1985), Keynes, prediction and econometrics, in Lawson and Pesaran (1985), 116–33.

(1987), The relative/absolute nature of knowledge, *Economic Journal* 97, 951–70.

(1989), Realism and instrumentalism in the development of econometrics, *Oxford Economic Papers* 41, 236–58.
Lawson, Tony and Hashem Pesaran, eds. (1985), *Keynes' Economics, Methodological Issues*, Routledge, London.
Leamer, Edward E. (1978), *Specification Searches, Ad Hoc Inference With Nonexperimental Data*, John Wiley, New York.
(1983), Let's take the con out of econometrics, *American Economic Review* 73, 31–43.
(1984), *Sources of International Comparative Advantage*, The MIT Press, Cambridge, MA.
(1985), Vector autoregressions for causal inference? In Karl Brunner and Allan H. Meltzer, eds. (1985), *Understanding Monetary Regimes*, *Carnegie-Rochester Conference Series on Public Policy* 22, 255–304, North Holland, Amsterdam.
(1989), Planning, criticism and revision, *Journal of Applied Econometrics* 4, S5–S27.
Lehmann, E. L. ([1959] 1986), *Testing Statistical Hypotheses*, John Wiley, New York.
Levi, Isaac (1967), *Gambling with Truth: an Essay on Induction and the Aims of Science,* Knopf, New York.
Levine, Ross and David Renelt (1992), A sensitivity analysis of cross-country growth regressions, *American Economic Review* 82, 942–63.
Li, Ming and Paul M. B. Vitányi (1990), Kolmogorov complexity and its applications, in J. van Leeuwen, ed. (1990), *Handbook of Theoretical Computer Science*, Elsevier Science Publishers, Amsterdam.
(1992), Inductive reasoning and Kolmogorov complexity, *Journal of Computer and System Sciences* 44, 343–84.
Liu, Ta-Chung (1960), Underidentification, structural estimation, and forecasting, *Econometrica* 28, 855–65.
Lucas, Robert E., Jr. (1980), Methods and problems in business cycle theory, *Journal of Money, Credit and Banking* 12, 696–715 (reprinted in Lucas, 1981).
(1981), *Studies in Business-Cycle Theory*, The MIT Press, Cambridge, MA.
Maasoumi, Esfandiar (1987), Information theory, in Eatwell, Milgate and Newman (1987), 846–51.
Mackie, J. L. I. (1965), Causes and conditions, *American Philosophical Quarterly* 2, 245–64.
MacKinnon, James G. (1983), Model specification tests against non-nested alternatives, *Econometric Reviews* 2, 85–158 (with discussion).
Maddala, G. S. (1977), *Econometrics*, McGraw-Hill, Auckland.
Marschak, Jacob (1941), A discussion of methods in economics, *Journal of Political Economy* 49, 441–8.
(1943), Money illusion and demand analysis, *Review of Economic Statistics* 25, 40–8.
McAleer, Michael, Adrian R. Pagan and Paul A. Volcker (1985), What will take the con out of econometrics? *American Economic Review* 75, 293–307.

McCloskey, Donald N. (1985), *The Rhetoric of Economics*, University of Wisconsin Press, Madison, WI.

Mill, John Stuart ([1843] 1952), *A System of Logic*, Longmans Green, London.

Mills, Frederick C. (1924), *Statistical Methods Applied to Economics and Business*, Henry Holt, New York.

(1938), *Statistical Methods Applied to Economics and Business*, revised edition, Henry Holt, New York.

Mirowski, Philip (1989), The probabilistic counterrevolution, or how stochastic concepts came to neoclassical economic theory, *Oxford Economic Papers* 41, 217–35.

(1995), Three ways to think about testing in econometrics, *Journal of Econometrics* 67, 25–46.

Mises, Richard von ([1928] 1981), *Probability, Statistics and Truth*, Dover, New York.

(1939), *Kleines Lehrbuch des Positivismus, Einführung in die empiristische Wissenschaftsauffassung*, W.P. van Stockum, Den Haag.

(1964), *Mathematical Theory of Probability and Statistics*, Academic Press, New York.

Mitchell, Wesley (1913), *Business Cycles*, University of California Press, Berkeley, CA.

Mizon, Grayham E. and Jean François Richard (1986), The encompassing principle and its application to testing non-nested hypotheses, *Econometrica* 54, 657–78.

Mood, Alexander M., Franklin A. Graybill and Duane C. Boes (1974), *Introduction to the Theory of Statistics*, McGraw Hill, Auckland.

Moore, Henry L. (1914), *Economic Cycles: Their Law and Cause*, Macmillan, New York.

Morgan, Mary S. (1990), *The History of Econometric Ideas*, Cambridge University Press.

Nagel, Ernest (1952), Review of Richard von Mises' *Wahrscheinlichkeit Statistik und Wahrheit, Econometrica* 20, 693–5.

Neumann, John von and Oskar Morgenstern, ([1944] 1947), *Theory of Games and Economic Behavior,* Princeton University Press.

Neyman, Jerzy (1952), *Lectures and Conferences on Mathematical Statistics and Probability*, Graduate School US Department of Agriculture, Washington.

(1977), Frequentist probability and frequentist statistics, *Synthese* 36, 97–131.

Neyman, Jerzy and Egon S. Pearson (1928), On the use and the interpretation of certain test criteria for purposes of statistical inference, *Biometrika* 20, 175–240 (Part I) and 263–94 (Part II), reprinted in Neyman and Pearson (1967).

(1933a), On the problem of the most efficient tests of statistical hypotheses, *Philosophical Transactions of the Royal Society* A 231, 289–337, reprinted in Neyman and Pearson 1967, 140–85.

(1933b), The testing of statistical hypotheses in relation to probabilities a priori, *Proceedings of the Cambridge Philosophical Society* 29, 492–510, reprinted in Neyman and Pearson 1967, 186–202.
(1967), *Joint Statistical Papers*, Cambridge University Press.
Nowak, Leszek (1980), *The Structure of Idealization*, Reidel, Dordrecht.
Pagan, Adrian R. (1995), On calibration, paper presented at the Seventh World Congress of the Econometric Society in Tokyo.
Palm, Franz and Arnold Zellner (1974), Time series analysis of simultaneous equation models, *Journal of Econometrics*, 17–54.
Pearson, Karl ([1892] 1911), *The Grammar of Science*, Adam & Charles Black, London.
(1900), On the criterion that a given system of deviations from the probable in the case of a correlated system of variables is such that it can be reasonably supposed to have arisen from random sampling, *Philosophical Magazine* 50, 157–72.
Pearson, Karl, ed. (1978), *The History of Statistics in the Seventeenth and Eighteenth Centuries*, Charles Griffin and Company Ltd., London.
Peirce, Charles S. (1892), The doctrine of necessity examined, *The Monist*, 321–37, reprinted in Peirce (1958), 160–79.
(1955), *Philosophical Writings of Peirce*, Dover Publications, New York.
(1958), *Selected Writings* (*Values in a Universe of Chance*), Dover Publications, New York.
Pesaran, M. Hashem (1982), On the comprehensive method of testing non-nested regression models, *Journal of Econometrics* 18, 263–74.
Phlips, Louis (1974), *Applied Consumption Analysis*, North Holland, Amsterdam.
Poincaré, J. Henri ([1903] 1952), *Science and Hypothesis*, English translation of *La Science et l'Hypothèse*, Dover, New York.
Pollock, Stephen S. G. (1992), Lagged dependent variables, distributed lags and autoregressive residuals, *Annales d'Économie et de Statistique* 28, 143–64.
Popper, Karl R. ([1935] 1968), *The Logic of Scientific Discovery*, Harper and Row, New York.
([1957] 1961), *The Poverty of Historicism*, Routledge and Kegan Paul, London.
([1963] 1989), *Conjectures and Refutations*, Routledge and Kegan Paul, London.
(1976), *Unended Quest, an Intellectual Autobiography*, Fontana/Collins, London.
(1982), *Quantum Theory and the Schism in Physics (from the postscript to the Logic of Scientific Discovery)*, Rowman and Littlefeld, London.
(1983), *Realism and the Aim of Science (from the postscript to the Logic of Scientific Discovery)*, Rowman and Littlefeld, London.
Pratt, John W. and Robert Schlaifer (1988), On the interpretation and observation of laws, *Journal of Econometrics* 39, 23–52.
Prescott, Edward E. (1986), Theory ahead of business cycle measurement, *Federal Reserve Bank of Minneapolis Quarterly Review* 10, 9–22.

Pudney, Stephen (1989), *Modelling Individual Choice, the Econometrics of Corners, Kinks and Holes*, Blackwell, Cambridge.
Putnam, Hilary (1963), 'Degree of confirmation' and inductive logic, in Schilpp (1963), 761–84.
Quah, Danny (1993), Galton's fallacy and tests of the convergence hypothesis, *Scandinavian Journal of Economics* 95, 427–43.
Quine, Willard van Orman ([1953] 1961), *From a Logical Point of View, Nine Logico-philosophical Essays*, Harvard University Press, Cambridge, MA.
(1987), *Quiddities*, Harvard University Press, Cambridge, MA.
Ramsey, Frank P. (1926), Truth and probability, in his *Foundations, Essays in Philosophy, Logic, Mathematics and Economics*, Routledge, London, 1978, 58–100.
(1929), General propositions and causality, in his *Foundations, Essays in Philosophy, Logic, Mathematics and Economics*, Routledge, London, 1978, 133–51.
Ramsey, James B. (1974), Classical model selection through specification error tests, in Zarembka (1974), 13–47.
Rao, C. R. ([1965] 1973), *Linear Statistical Inference and its Applications*, John Wiley, New York.
Reichenbach, Hans (1935), *Wahrscheinlichkeitslehre*, Sijthoff, Leiden.
([1938] 1976), *Experience and Prediction, an Analysis of the Foundation and the Structure of Knowledge*, University of Chicago Press, Midway reprint, Chicago, IL.
Rhodes, Richard (1986), *The Making of the Atomic Bomb*, Penguin Books, London.
Rissanen, Jorma (1983), A universal prior for integers and estimation by minimum description length, *Annals of Statistics* 11, 416–31.
(1987), Stochastic complexity and the MDL principle, *Econometric Reviews* 6, 85–102.
Robbins, Lionel ([1932] 1935), *An Essay on the Nature & Significance of Economic Science*, Macmillan, London.
Rosenkrantz, Roger (1983), Why Glymour is a Bayesian, in Earman (1983), 69–97.
Russell, Bertrand (1913), On the notion of cause, *Proceedings of the Aristotelian Society* 13, 1–26.
(1927), *An Outline of Philosophy*, Allen & Unwin, London.
(1946), *A History of Western Philosophy*, Allen & Unwin, London.
Samuelson, Paul A. ([1947] 1983), *Foundations of Economic Analysis*, Harvard University Press, Cambridge, MA.
Sargent, Thomas J. (1981), Interpreting economic time series, *Journal of Political Economy* 89, 213–48.
Savage, Leonard J. (1952), Review of Rudolf Carnap's *Logical Foundations of Probability*, *Econometrica* 20, 688–90.
([1954] 1972), *The Foundations of Statistics*, Dover Publications, New York.

(1961), The foundations of statistics reconsidered, *Proceedings of the Fourth Berkeley Symposium on Mathematical Statistics and Probability* 1, ed. Jerzy Neyman, University of California Press, Berkeley, CA.

(1962), Subjective probability and statistical practice, in G. A. Barnard and D. A. Cox, eds. (1962), *The Foundations of Statistical Inference*, Methuen, London. 9–35.

Sawa, Takamitsu (1978), Information criteria for discriminating among alternative regression models, *Econometrica* 46, 1273–91.

Schilpp, Paul Arthur, ed. (1963), *The Philosophy of Rudolf Carnap*, Open Court, La Salle, IL.

Schultz, Henry (1928), *Statistical Laws of Demand and Supply (with special applications to sugar)*, University of Chicago Press, Chicago, IL.

(1937), *The Theory and Measurement of Demand*, University of Chicago Press, Chicago, IL.

Schwarz, G. (1978), Estimating the dimension of a model, *Annals of Statistics* 6, 461–4.

Sheenan, R. G. and R. Grives (1982), Sunspots and cycles: a test of causation, *Southern Economic Journal* 48, 775–7.

Simon, Herbert A. (1953), Causal ordering and identifiability, in Wm. C. Hood and Tjalling C. Koopmans, *Studies in Econometric Methods,* Wiley, New York, 49–74.

(1986), Rationality in psychology and economics, in Hogarth and Reder (1986), 25–40.

Sims, Christopher (1972), Money, income and causality, *American Economic Review* 62, 540–52.

(1980), Macroeconomics and reality, *Econometrica* 48, 1–48.

(1996), Macroeconomics and methodology, *Journal of Economic Perspectives* 10, 105–20.

Sims, Christopher and Harald Uhlig (1991), Understanding unit rooters: a helicopter tour, *Econometrica* 59, 1591–9.

Skidelski, Robert (1983), *John Maynard Keynes, Hopes Betrayed 1883–1920*, Viking, New York.

Solomonoff, Ray (1964a), A formal theory of inductive inference, I, *Information and Control* 7, 1–22.

(1964b), A formal theory of inductive inference, II, *Information and Control* 7, 224–54.

Spanos, Aris (1986), *Statistical Foundations of Econometric Modelling*, Cambridge University Press.

(1989), On rereading Haavelmo: A retrospective view of econometric modeling, *Econometric Theory* 5, 405–29.

Stigler, George J. (1962), Henry L. Moore and statistical economics, *Econometrica* 30, 1–21.

(1965), *Essays in the History of Economics*, University of Chicago Press, Chicago, IL.

Stigler, Stephen M. (1986), *The History of Statistics, the Measurement of Uncertainty before 1900*, Harvard University Press, Cambridge, MA.

(1987), The measurement of uncertainty in nineteenth-century social science, in Krüger, Daston and Heidelberger (1987), 287–92.

Stigum, Bernt P. (1990), *Toward a Formal Science of Economics*, The MIT Press, Cambridge, MA.

Stone, J. N. R. (1954a), *The Measurement of Consumer's Expenditure and Behaviour in the United Kingdom, 1920–1938*, volume I, Cambridge University Press.

(1954b), Linear expenditure systems and demand analysis: an application to the pattern of British demand, *Economic Journal* 64, 511–27.

Stouffer, A. (1934), Sociology and sampling, in L. L. Bernard (ed.), *Fields and Methods of Sociology*, Farrar & Rinehart, New York, 476–87.

Strotz, Robert H. and H. O. A. Wold (1960), Recursive vs nonrecursive systems: an attempt at synthesis, *Econometrica* 28, 417–27.

Summers, Lawrence (1991), The scientific illusion in empirical macroeconomics, *Scandinavian Journal of Economics* 93, 129–48.

Suppes, Patrick (1970), *A Probabilistic Theory of Causality*, North Holland, Amsterdam.

Swamy, P. A. V. B., R. K. Conway and P. von zur Muehlen (1985), The foundations of econometrics, are there any?, *Econometric Reviews* 4, 1–61.

Theil, Henri (1965), The information approach to demand analysis, *Econometrica* 33, 67–87.

(1971), *Principles of Econometrics*, John Wiley, New York.

(1980), *The System-Wide Approach to Microeconomics*, University of Chicago Press, Chicago, IL.

Thorburn, W. M. (1918), The myth of Occam's razor, *Mind* 27, 345–53.

Tinbergen, Jan (1927), Over de mathematies-statistiese methoden voor konjunktuuronderzoek, *De Economist* 76, 711–23.

(1936), Prae-advies, Vereeniging voor de Staatshuishoudkunde en de Statistiek, Nijhoff, The Hague.

(1939a), *Statistical Testing of Business Cycle Theories*, I: *A Method and its Application to Investment Activity*, League of Nations Economic Intelligence Service, Geneva.

(1939b), *Statistical Testing of Business Cycle Theories*, II: *Business Cycles in the United States of America*, League of Nations Economic Intelligence Service, Geneva.

(1940), On a method of statistical business-cycle research. A reply, *Economic Journal* 50, 141–54.

Turner, Stephen P. (1986), *The Search for a Methodology of Social Science*, Reidel Publishing Company, Dordrecht.

Tversky, Amos and Daniel Kahneman (1986), Rational choice and the framing of decisions, in Hogarth and Reder (1986), 67–74.

Varian, Hal (1983), Nonparametric tests of consumer behaviour, *Review of Economic Studies* 50, 99–110.

(1990), Goodness-of-fit in optimizing models, *Journal of Econometrics* 46, 125–40.
Venn, John ([1866] 1876), *The Logic of Chance: an Essay on the Foundations and Province of the Theory of Probability, with Especial Reference to its Logical Bearings and its Application to Moral and Social Science*, Macmillan, London.
Vercelli, Alessandro and Nicola Dimitri, eds. (1992), *Macroeconomics, a Survey of Research Strategies*, Oxford University Press.
Vining, Rutledge (1949), Koopmans on the choice of variables to be studied and of methods of measurement, *Review of Economic Statistics* 31, 77-86; 91–4.
Vuong, Quang H. (1989), Likelihood ratio tests for model selection and non-nested hypotheses, *Econometrica* 57, 307–33.
Wald, Abraham (1940), Über die Wiederspruchsfreiheit des Kollektivbegriffes, *Ergebnisse eines mathematischen Kolloquiums*, No. 8, Vienna, 38–72.
Walras, Leon ([1874–7] 1954), *Elements of Pure Economics*, Irwin, Homewood, IL.
Watkins, John (1984), *Science and Scepticism*, Hutchinson, London.
Whittle, Peter ([1963] 1983), *Prediction and Regulation by Linear Least Squares Methods*, University of Minnesota Press, Minneapolis, MN.
Wiener, Norbert ([1948] 1961), *Cybernetics, or the Control and Communication in the Animal and the Machine*, The MIT Press, Cambridge, MA.
(1956), The theory of prediction, in E. F. Beckenback, ed. (1956), *Modern Mathematics for Engineers*, McGraw Hill, New York, 165–90.
Wold, Herman A., in association with L. Juréen (1953), *Demand Analysis*, John Wiley, New York.
Wrinch, Dorothy and Harold Jeffreys (1921), On certain fundamental principles of scientific inquiry, *The London, Edinburgh and Dublin Philosophical Magazine and Journal of Science* 42, 369–90.
Yule, George Udny ([1911] 1929), *An Introduction to the Theory of Statistics*, Charles Griffin and Company, London.
(1921), On the time-correlation problem, with especial reference to the variate-difference correlation method, *Journal of the Royal Statistical Society* 84, 497–526.
(1926), Why do we sometimes get nonsense-correlations between time-series? – a study in sampling and the nature of time-series, *Journal of the Royal Statistical Society* 89, 1–64.
Zarembka, Paul, ed. (1974), *Frontiers in Econometrics*, Academic Press, New York.
Zellner, Arnold (1971), *An Introduction to Bayesian Inference in Econometrics*, Wiley, New York.
(1978), Jeffreys–Bayes posterior odds ratio and the Akaike information criterion for discriminating between models, *Economic Letters* 1, 337–42.
(1979), Causality and econometrics, in K. Brunner and A. H. Meltzer, eds.

(1979), *Three Aspects of Policy and Policy Making: Knowledge, Data and Institutions*, North Holland, Amsterdam, reprinted in Zellner (1984), 35–74.
(1984), *Basic Issues in Econometrics*, University of Chicago Press, Chicago, IL.
(1988), Causality and causal laws in econometrics, *Journal of Econometrics* 39, 7–21.

Name Index

Subject index